PLANCK

PLANCK

DRIVEN BY VISION, BROKEN BY WAR

BRANDON R. BROWN

OXFORD
UNIVERSITY PRESS

Oxford University Press is a department of the University of Oxford. It furthers the University's objective of excellence in research, scholarship, and education by publishing worldwide.

Oxford New York
Auckland Cape Town Dar es Salaam Hong Kong Karachi
Kuala Lumpur Madrid Melbourne Mexico City Nairobi
New Delhi Shanghai Taipei Toronto

With offices in
Argentina Austria Brazil Chile Czech Republic France Greece
Guatemala Hungary Italy Japan Poland Portugal Singapore
South Korea Switzerland Thailand Turkey Ukraine Vietnam

Published in the United States of America by
Oxford University Press
198 Madison Avenue, New York, NY 10016

Cover photo courtesy AIP Emilio Segre Visual Archives.

Library of Congress Cataloging-in-Publication Data
Brown, Brandon R.
Planck : driven by vision, broken by war / Brandon R. Brown.
pages cm
Includes bibliographical references and index.
ISBN 978-0-19-021947-5 (hardback : alkaline paper) 1. Planck, Max, 1858–1947. 2. Physicists—Germany—Biography. 3. World War, 1939–1945—Science—Germany. 4. National socialism and science. 5. Planck, Max, 1858–1947—Family. 6. Planck, Max, 1858–1947—Friends and associates. 7. Einstein, Albert, 1879–1955. 8. Physics—Germany—History—20th century. 9. Science—Germany—History—20th century. 10. Germany—Intellectual life—20th century. I. Title.
QC16.P6B76 2015
530.092—dc23
[B]
2014042822

1 3 5 7 9 8 6 4 2
Printed in the United States of America
on acid-free paper

To the writer who told me, "Stick to physics, and write from there."

Contents

Preface

Science has often found unique ways to humiliate its devotees. In 1964, two youngish men crawled inside an enormous metal basket carrying brushes and a bucket of soapy water. The 20-foot-long radio receiver looked like the head of a lacrosse stick, but functioned like an ear horn, opening to the heavens and listening to the cosmos. They scrubbed and scrubbed, hoping the caked layer of pigeon poop might (gloved fingers crossed) be the cause of the mysterious, frustrating, and unwanted signal. Maybe taking a manual Q-tip to the horn antenna's ear would clear up the one note of noise.

They were an unlikely pair in an unlikely place. Robert Woodrow Wilson was a Houston native, and Arno Penzias was a German immigrant who had escaped one of Hitler's camps at age six. And this hill overlooking New York City was not a normal place to pursue astronomy. Bell Labs had designed the antenna to communicate with the new Telstar satellite, but with down time from its primary duties, the owners let astronomers give it a whirl. Wilson and Penzias wanted to probe the sparse outer reaches of the Milky Way galaxy.[1]

They needed the signal to be pristine, and with great effort, they had fine-tuned and calibrated the big basket for their measurements. They found ways to filter out local radio broadcasts, noisy radar echoes, and other extraneous signals arising from their own electronics. After all this, the horn still had some kind of tinnitus—there was a little ringing at the wavelength 7.35 centimeters. No matter where in the cosmos they pointed the antenna, no matter the time of day or night, there was the same ringing, always the same strength. The only thing that all directions and all times of day had in common, they figured, was bird crap. When they ran the big device on chilly nights, pigeons gathered at the warm end and made a mess.

Bird crap removed, they pointed the instrument again away from the thick plane of the Milky Way galaxy and out into the darkest, deepest reaches of space. They wanted to make sure the signal was gone, like listening for an unwanted hum in an excellent audio system. Just sit still in the dark until the speakers emitted only pristine, quiet beauty. But no, the ghastly noise peak was there again, as strong as ever. The universe appeared to be emitting radiation similar to that from the Amana Corporation's new microwave ovens. With heavy sighs, Wilson and Penzias recorded and annotated the mysterious tone, just in case it wasn't a fingerprint of their own incompetence.

The two had unwittingly made a critical discovery, one for which they would win the Nobel Prize in Physics: The microwave signal is a dim but very real glow emanating uniformly from the universe itself. In further measurements, the "cosmic microwave background" perfectly fit an equation dating to 1900, when the German physicist Max Planck described the natural radiation emitted by *any* object at *any* temperature, be it a bright fiery star, a lukewarm nickel in your pocket, or in the case of the universe's background signal—a faint afterglow of the Big Bang (Figure P.1).[2]

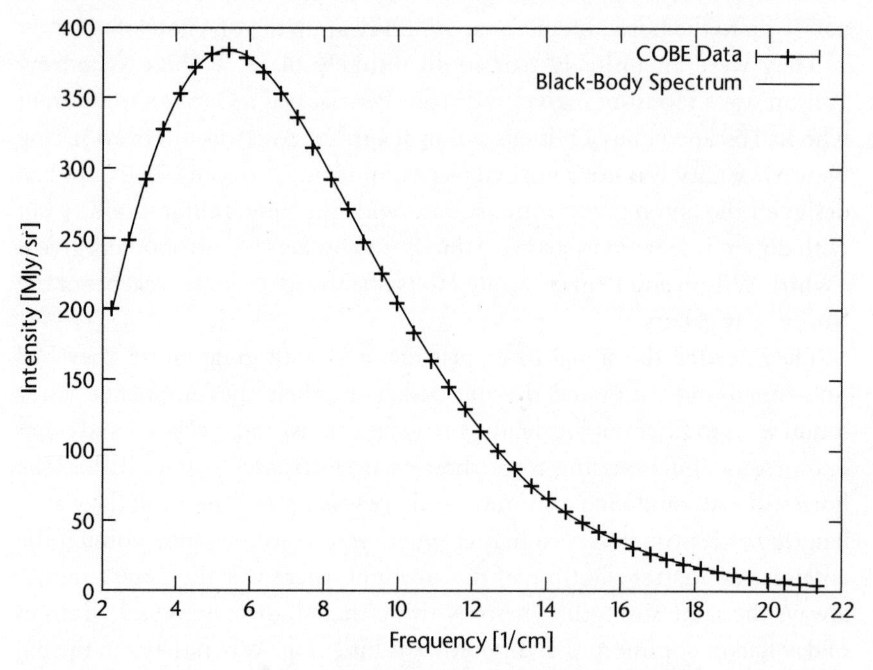

Figure P.1. Spectrum of the cosmic background radiation, or the residual glow of the universe, when all galaxies, dust, and so on are removed from the signal. When these data from the COBE satellite were presented in 1990, they precipitated a spontaneous standing ovation. Planck's law (*the line labeled Black-Body Spectrum*) fits the universe's signal (*crosses*) so precisely that the uncertainties in the data points are much smaller than the line thickness used here.

As of this writing, humanity now has a cosmic ear horn in orbit, and it listens with the best clarity yet to these low-frequency signals from the universe. Similar in size and shape to the instrument used by Wilson and Penzias, it rotates its narrow view about once every minute, sweeping out a ring of measurements like a second hand. In the cosmic background's blemishes and tiny inconsistencies, the Planck satellite can see residual clues describing the universe's initial fireball—a fingerprint of the *first* physics, or the very hands of God, depending on whom you ask. The satellite's namesake, the late German physicist Max Planck, did not think much about astronomy, and when his younger friend Albert Einstein turned his own gaze to the cosmos, Planck told him it was probably a waste of time. Yet, when the European Space Agency decided it needed a catchier project title than the acronym COBRAS/SAMBA, the name "Planck" was an easy sell to all parties.

It had been just as easy to sell his name after World War II, as the Allies looked to rebrand all German research programs. Albert Einstein, estranged from Planck and bitterly divorced from Germany, composed a tribute to the man on behalf of American scientists. "Even in these times of ours," he wrote in 1948, "when political passion and brute force hang like swords over the anguished heads of men, that even in such times there is being held high and undimmed the standard of our ideal search for truth. This ideal, a bond forever uniting scientists in all times and in all places, was realized with rare completeness in Max Planck." And according to Einstein, Planck's 1900 discovery, "became the basis of all 20th century research in physics and has almost entirely conditioned its development ever since. Without this discovery it would not have been possible to establish a workable theory of atoms and molecules and the energetic processes which govern their transformations."[3] This was not a hyperbolic statement then, and it holds today.

Our understanding of the building blocks and the structure of matter trace directly to Planck's work. And our understanding of how separate chunks of matter then exchange energy—how they chat and inform one another—also starts with Planck's primary discovery. He expertly described the radiation that leaks from any and every object in the universe. No matter what object, and no matter its temperature, we need just one equation—Planck's—to describe every single case. At the time he penned his formula, scientists were years from discovering galaxies beyond our own, never mind looking for remnants of the Big Bang. Planck, just like Wilson and Penzias, had been trying to diagnose one thing when he stubbed his toe on something much different and even more important. In trying to once and for all describe this baffling glow from all things—called "black-body radiation"—Planck found the key that unlocked the modern age of physics. Even though he contemplated the physics governing the light *inside* of a small, dark cavity within a brick, his satellite now gazes in the opposite direction—the ultimate outward—and finds the same fundamental physical law reigning supreme.

Planck is known as the father of quantum theory, and most textbooks give students little more than that. He was German. He was a theoretical physicist (versus an experimental, or laboratory-based one), with a firm grasp of mathematics. In the typical side-column photo, we see him later in life: bald, and stern. He discovered quantum theory. He had a mustache. And that's about it (Figure P.2).

But there is so much more to Planck the scientist and Planck the person.

Figure P.2. Max Planck in 1906, at age 48.
Photograph by Rudolf Dührkoop, courtesy AIP Emilio Segre Visual Archives, W. F. Meggers Gallery of Nobel Laureates.

Max Planck had elevated and refined the formerly obscure notion of "entropy" in the universe—he made it not only a useful tool, but also a central topic. Relevant for diagnostics ranging from car engines to black holes, entropy has even provided a template for the study of information itself. Planck also made great contributions to chemistry, to the then-infant field of statistical mechanics, and to Albert Einstein's new ideas of relativity.

His human story is equally rich: musical ability, a cherished family, and a sterling reputation; a devotion to his homeland, come what may; a delicate and poignant relationship with Albert Einstein. Planck was first and last a communicator. He assembled prose in the manner of a master watchmaker, and he launched his mind at much more than physics. Planck was also a person in the right place at all the wrong times, watching ridiculous advances in technology reformat his world and then tear it apart. In 1933, just as little Arno Penzias was born into a nervous German Jewish family, Planck was trying to reason with the new German Chancellor Adolf Hitler.

After Planck's death, the Royal Society sent Charles Darwin's grandson, Charles George Darwin, to Berlin. Even though the Britain of 1948 had no love lost for Germany, one name transcended the garish wounds of two wars. "But if Planck the originator in scientific achievement commands the homage of our heads," Darwin said, "no less does Planck the man deserve the approbation of our hearts. His character was modest, kindly and blameless, and amid the trials of distressful times and through many personal sorrows he preserved his integrity and his quiet courage."

There are many sensible reasons that Planck's story is not better known, particularly in the English language. His library, personal journals, notebooks, and letters were destroyed with his home in World War II. What exists of his correspondence with other German scientists is often handwritten in an antiquated form of German shorthand, *Sütterlin*, understood by ever-fewer scholars. And he was certainly eclipsed by the younger, bolder, and more brilliant Albert Einstein. Whereas Planck was very much a nineteenth-century Prussian gentleman walking into a wholly new twentieth century, Einstein saw himself as a modern man of the world, and he benefited from the dawn of global media. He also enjoyed a long presence in America as it took the mantle of worldwide scientific leadership from Planck's vanquished Germany.

Humbly, I now try to tell some of Max Planck's rich story. I admit from the start that I can't approach his life as a science historian, but I come to Planck as a physicist long fascinated by his breakthrough and haunted by those sad eyes. I have for many years wanted to know who he was, what shaped him, and how we might best understand his circumstances—or, as we might say in physics, his fundamental principles, his initial conditions, and his boundary conditions. What follows are my best attempts to discover this German physicist and share the results—not just with scientists, but with any interested reader, since we are bathed one and all, from every direction, in the glow of his law.

Brandon R. Brown, Summer, 2014.

Acknowledgments

I humbly offer my gratitude to Dr. Dieter Hoffmann for his helpful advice and communication. May American academics learn something not only from his decades of scholarship on German science, but also from his hospitality to a nonspecialist.

For enormous language and cultural translation help, I thank Dr. Imke Listerman and Candice Novak. (Any residual translation errors are without a doubt mine alone.)

For generous, enjoyable, and immensely helpful discussions on many technical topics in this book, I thank Dr. Horacio Camblong.

For the sabbatical of 2012–2013, I thank my Dean's office, as well as my faculty union, the USFFA. I thank the University of San Francisco's faculty development fund for supporting international travel, and translation fees. I also thank my home department for great support and camaraderie throughout this project.

I acknowledge the cooperation and assistance of the following archives and institutes: Archiv der Max-Planck-Gesellschaft in Berlin-Dahlem, with special thanks to Dr. Kristina Starkloff, Dirk Ullmann, Susanne Uebele, and Bernd Hoffmann; the Deutsches Museum Archiv in Munich, with special thanks to Dr. Matthias Röschner; the Archives of the American Philosophical Society in Philadelphia, with special thanks to Earle Spamer; the American Institute of Physics, with special thanks to Melanie Mueller; the Emilio Segre Visual Archives, with special thanks to Savannah Gignac; the Niels Bohr Library, with special thanks to Felicity Pors; Brookhaven National Laboratory, with special thanks to Jane Koropsak; Stadtmuseum Kassel, with special thanks to Dr. Alexander Link; and finally Germany's Bundesarchiv, with special thanks to Sabrina Bader.

For helpful conversations and good ideas along the way, I thank Morgen Daniels, Dr. Bob Eason, Dr. Michael Eckert, Arden Hendrie, Peter Hennings, Victor Lin, Dr. Elliot Neaman, Spin City Coffee, Dr. Dean

Rader, Rachel Reed, Dr. Jennifer Turpin, Dr. Aparna Venkatesan, Pete Zuraw, and Terry Zuraw.

I thank my agent Jennifer Lyons for her advice and advocacy.

I thank my editor Jeremy Lewis and his great team at Oxford University Press for guidance, enthusiasm, and clarity.

I thank my dozens of teachers, professors, and colleagues who sparked and honed interest. For writing training, I thank in chronological order: Mom, Dad, Mrs. Gohmert, Ms. Miller, Max Apple, Glen Blake, Ehud Havezelet, and Rich Daniels. For years of writing instruction and special advice for this project, I also thank Marjorie Sandor and Tracy Daugherty. For reporting training, I want to thank Johnny Mac and all my fellow Santa Cruz slugs.

Finally, I thank my wife, Dana Smith, for years of brainstorming, months of proofreading, and endless patience with my Planck obsessions.

Max Planck Timeline

1858	Karl Ernst Ludwig Marx Planck (M.P.) born in Kiel, Denmark.
1867	Planck family moves from Kiel to Munich, Bavaria.
1870	M.P.'s older brother Hermann dies in the Franco-Prussian War.
1871	Wilhelm I of Prussia proclaimed Emperor of newly formed Germany.
1874	M.P. begins studies at the University of Munich.
1879	M.P. defends dissertation and notes entropy as the arrow of time.
1880	M.P. defends habilitation thesis.
1885	M.P. appointed associate professor of theoretical physics, University of Kiel.
1887	M.P. weds Marie Merck in Munich.
1888	Karl Planck, first child of M.P. and Marie, is born.
1889	M.P. accepts position at the University of Berlin. Twins Emma and Greta Planck are born.
1893	Erwin Planck is born, as the last child of M.P. and Marie.
1894	M.P. elected to Berlin Academy of Sciences turns attention to black-body radiation.
1900	M.P.'s father Wilhelm dies in Munich. M.P. effectively launches quantum theory during brief presentation in Berlin.
1903	Max Laue obtains PhD under thesis advisor M.P.
1905	The Plancks move to house in Grunewald suburb of Berlin. M.P. elected president of Prussian Physical Society. "Miracle year" for Albert Einstein, with five groundbreaking publications.
1906–1908	M.P. publishes relativistic dynamics, building on Einstein's initial work.
1907	Austrian student Lise Meitner comes to Berlin to attend M.P. lectures.
1909	M.P. visits the United States and gives eight lectures at Columbia University. On his return, M.P.'s first wife Marie dies. At a conference in Salzburg, M.P. meets Albert Einstein for the first time.
1911	M.P. marries Marga von Hoesslin. First meeting on quantum theory at the Solvay conference, in Brussels. M.P. provides clear

	statement of Third Law of Thermodynamics. Hermann Planck born to Max and Marga Planck.
1911–1913	M.P. introduces a "zero point" energy of the vacuum in his second theory of thermal radiation.
1912	M.P. appoints Lise Meitner as his new assistant. Karl Planck is admitted to a sanatorium in Kassel.
1913	M.P. becomes rector for the University of Berlin and joins the Kaiser Wilhelm Society (KWG). Albert Einstein officially accepts academic post in Berlin. Grete Planck is admitted to a sanatorium.
1914	World War I begins. M.P.'s mother Emma dies in Munich at age 93. M.P. signs "Manifesto of the 93 Intellectuals" defending Germany. Erwin Planck is taken prisoner by French forces.
1916	Karl Planck dies in the battle of Verdun.
1917	Grete Planck dies shortly after childbirth; her daughter Grete Marie survives. Erwin Planck is released from captivity in France and returns to Berlin.
1918	Kaiser Wilhelm II abdicates his throne, leading to end of World War I.
1919	M.P. awarded 1918 Nobel Prize in Physics. Emma Planck dies in childbirth; her daughter Emmerle survives.
1920	M.P. and colleagues launch *Notgemeinschaft* to fund German science.
1923	Attempted Nazi coup fails in Munich. Erwin Planck marries Nelly Schoeller.
1926	M.P. officially retires as a professor (duties continue for some time).
1930	M.P. becomes president of the KWG.
1933	Adolf Hitler is appointed Chancellor, and Erwin Planck resigns from government. M.P. has face-to-face meeting with Hitler.
1937	M.P. steps down from presidency of the KWG.
1938	M.P. is forced from presidency of the Prussian Academy of Science.
1939	Germany invades Poland, officially starting World War II.
1943	M.P. and Marga move to Rogätz due to ongoing bombing in Berlin.
1944	Adolf Hitler survives assassination attempt. Erwin Planck is arrested in connection and found guilty of treason.
1945	Erwin Planck is executed at Plötzensee Prison. M.P. and Marga are rescued by the American Alsos Mission.
1947	Karl Ernst Ludwig Marx Planck dies at age 89 in Göttingen.

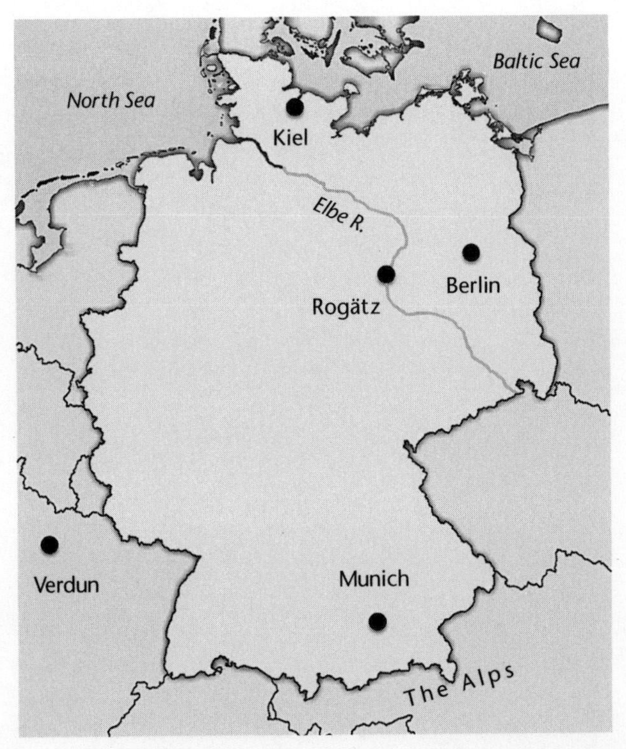

This map of modern Germany highlights key locations and features for the life of Max Planck (1858–1947), while omitting many other features. The modern nations bordering Germany are, moving clockwise from the top: Denmark, Poland, Czech Republic, Austria, Switzerland, France, Luxembourg, Belgium, and the Netherlands.

PLANCK

I

October 1944

When Max Planck heard his son's verdict, he addressed a letter to Adolf Hitler. He wrote with calm respect and a surgeon's precision, as always, but he conveyed his shock. And the man of physics bartered as follows: If I am a national treasure, as you say, then show mercy. Reward this 87-year-old man's lifetime of work for the Fatherland by sparing my son's life.

Before this unique letter, Planck had spent decades methodically separating the worlds of work and family. He was a dignified Prussian gentleman, after all, so he didn't mention personal heartbreaks in his scientific correspondence. Eventually, some close friends believed he relied on work to escape the weight of family tragedy. But by 1944, as an avid mountain climber long past his summit, Planck had lost his footing, and the letter was like a pick to the mountainside, his best shot to halt a steep slide.

At the time, he stood as one of the most decorated physicists on the planet. The name Planck was well on its way to physics immortality, and he attracted nearly reverential respect and praise from scientists everywhere. He had won a Nobel Prize for unleashing the revolution of quantum theory. As he put his physics calculations to the side late in life, he emerged as a scientific philosopher and spokesman, with frequent public talks and radio interviews in peacetime and in war.

But during World War II, Planck's beloved homeland had also come to tarnish him. As an old friend of Albert Einstein (the despicable king of "Jewish Physics" to some), Planck faced decorated enemies now in science, like fellow Nobel laureates Johannes Stark and Philipp Lenard, and within the Reich leadership, like Reich Minister of Propaganda Joseph Goebbels. The Gestapo investigated his heritage, and newspaper columns referred to him as a "white Jew," one of the pied pipers who had supposedly led German science astray, taking students and colleagues into the barren field of more mathematical and less meaningful science.[1] His questionable associations didn't end with Einstein either; Planck had befriended other Jews, like the chemist Fritz Haber and the nuclear physicist Lise Meitner, both of whom had fled Nazi Germany. Worst of all, he'd requested a face-to-face meeting with Hitler in the spring of 1933, during the new Chancellor's first months of power. Planck's perceived criticism sparked a notoriously angry and one-sided confrontation.[2]

Erwin Planck was Max's long-time hiking partner and his favorite child. As a high-ranking member of the German government preceding Hitler's ascent, Erwin was no fan of the Nazi regime. They had in some cases murdered his friends and former colleagues. Erwin resigned from government in 1933 and accepted a job with a leading German manufacturer, staying away from politics. But in the summer of 1944, a briefcase exploded in Hitler's inner sanctum, barely missing its mark. The wounded Führer took to the radio swearing vengeance, and the Gestapo arrested hundreds in the days that followed. They took Erwin on July 23, accusing him of high treason against the Fatherland. They linked him to the conspirators, and for three months, Erwin awaited trial in prison. His captors denied the visit requests of Erwin's wife Nelly and his father. As with most prisoners so accused, he suffered intense interrogation sessions and, most probably, torture.[3]

Though Erwin apparently had no direct involvement in the bombing, he had helped draft a secret constitution for a post-Nazi government. Under interrogation, he admitted knowing the conspirators but claimed to have ended communication with them years earlier. In truth, he had provided them contacts and recruited supporters.[4]

The trials spawned by the assassination attempt began in August. Max Planck would have followed the few that were publicized, via radio and newspaper—the regime never wanted the public to know the conspiracy's full measure.[5] In the first weeks after the July bombing, 110 death sentences—*Tod!*—fell on suspects.[6] In bridging Germany's proudest moments to the horror of its Nazi present, Max Planck would only need to place his father Wilhelm Planck at one end and his helpless son Erwin at the other. His father, a nineteenth-century judicial scholar, helped refine Germany's civil code, proudly building upon the rewards of the enlightenment. But a fascist regime now paraded Erwin into the notorious People's Court, with an all-powerful judge sneering at due process.

We have a photo of Erwin standing at trial (Figure 1.1). His look, with a slack face and eyes unfocused, says he knows the verdict before hearing it. Max and Nelly referred to Erwin affectionately as "Mops," a German term for pug dog. Despite a family built on hope, Erwin had little in that moment. It was October 23 when Roland Freisler, president and presiding judge of the People's Court, spit another guilty verdict and another sentence to death by hanging, this time for Erwin Planck, son of the nation's figurehead of science. And so the father's race to save the son began. As he later confided to a friend, he would set "Heaven and Hell in motion" to that end.[7] If the sentence could be commuted to life imprisonment, Erwin might survive to the war's end, and they might see one another again.

For now, if Erwin was treated like others with a death sentence, he was escorted into Plötzensee Prison, a large three-story cross of a building. Built northwest of Berlin in the late nineteenth century, the site now houses a memorial to the thousands murdered there. The majority of executions took place in the last years of World War II. Most victims at Plötzensee

Figure 1.1. Erwin Planck on trial in the People's Court, October 1944.
Courtesy Archiv der Max-Planck-Gesellschaft, Berlin-Dahlem.

were either foreign nationals or Germans, like Erwin and his boyhood friend and neighbor Ernst von Harnack, also implicated in the resistance.

Convicts typically entered House III, the cellblock adjacent to the execution chamber. For many years, a guillotine served as the primary instrument, but in 1942, the Reich installed a large steel beam against one wall of the eerily blank cubic space, and there they mounted a set of iron hooks. In this way, they could carry out eight executions at a time, sometimes using piano wire. Over 250 executions took place in September of 1944 alone, including suspected members of the German resistance, as well as many from the Czech resistance. And as with the court sessions, the executions were filmed for Hitler's later viewing, particularly for anyone suspected in the bomb plot.[8]

By 1944, Planck's inner circle had either fled Germany, or, not sharing Planck's incredible vigor, had passed away. Erwin was now his father's closest confidant and best friend. He was also the last survivor of Max's four children with his beloved first wife, Marie. Erwin represented the last glimpse of incredible days in the early century, when optimism wasn't just a guiding principle—it was an obvious conclusion.

In 1905, the Planck family, the Kaiser's empire, and Max's well-nurtured garden of German physics all looked poised for decades of health and growth. New ideas and technology bristled in Berlin like a spring burst of wildflowers: electricity, automobiles, radio waves, moving pictures, and on and on.

By that year, Planck had made his most critical contribution to science, though very few recognized it at the time, including Planck himself. He'd been drawn to hints of transcendent and universal principles underlying a mysterious and little-studied phenomenon known as "black-body radiation," a type of energy that emerges from within matter. If one removes the reflected light bouncing from an object and measures only that emerging from its interior, every object, regardless of shape, size, or material, throws off the same exact type of radiation. A train car, a puppy, and a straw hat, if all the same temperature, give off the same faint signature: the same exact outline of frequencies, mostly in the infrared. Max Planck wanted to know *why*, rightly intuiting that the answer would dwarf the question. To see the experimental data from this realm of thermal physics was like walking into a world where a breeze would set every object ringing with the tone of a single wind chime—buildings, street signs, and ham sandwiches all giving the same exact tone. After years of toil, Planck uncovered a new fundamental constant, *h*, the "quantum of action" as he came to call it. This constant unlocked the radiation's underlying mathematical machinery, and when Planck first tried to describe what he saw, he unknowingly sparked the quantum revolution.

But most exciting to Max himself, and of great relevance to physics still, he had proposed the idea of "natural units": a system of measurements based only on fundamental universal constants, with no bias from human preference, convenience, or experience. The length of an arm or the span of a heartbeat would have no say in this system. Natural units would be so objective, Max Planck said, that any group of scientists, not merely humans but even extraterrestrials, would necessarily arrive at these units and agree on their values. To this day, we still refer to the Planck time (5.39106×10^{-44} second) and the Planck length (1.616199×10^{-35} meter). These are absolute measures derived directly from the fabric of the cosmos, via fundamental constants. We compute them by multiplying and dividing different combinations of the speed of light *c*, the universal constant of gravitation *G*, and Planck's own *h*, until we respectively obtain measures of length or time.[9] (The Planck length and Planck time are unfathomably tiny and short. As the city of San Francisco dwarfs a single proton, so that proton then dwarfs the Planck length by the same factor.)

In fact, since Planck unearthed the values of multiple fundamental constants with unprecedented precision and accuracy, he propelled the entire field of physics into a more numerically precise age. Quantum field theory, the grandchild of Planck's quantum notion, is now the most physically accurate theory in the history of science; its precision is such that a quantum doctor would be able to predict the lifespan of a patient to within a fraction of one second, or a quantum economist would be able to predict next year's U.S. gross domestic product to within $50.

Planck had also helped cement Germany, if not Berlin itself, as the absolute center of physics on Earth. German was *the* required language for any student of the increasingly respected subject of physics, and the journal *Annalen der Physik*, where Max served as editor, was its leading voice. As electric power innervated the cities of Europe and radio messages took their maiden voyages across oceans in an instant, interest in physics soared, classrooms filled, and the brightest students from Europe (and America) queued to get in.

The center of Planck's personal universe in 1905 was a new suburban address: Wangenheimstrasse 21 in Grunewald (literally "green forest"), a

Berlin suburb. Neighbors reported a dark-paneled interior with down-to-earth furnishings—every detail spoke to a sober and strict atmosphere.[10] At this address, Max enjoyed his ideal study and library, a chamber for intensive mornings of thinking and writing. He established a focused routine of study, writing, and correspondence in the mornings, with walks and music in the afternoons, and time for family in the evenings. Though projecting an austere public presence, sometimes interpreted as aloof or arrogant, Planck could also be very playful within the walls of his villa on Wangenheimstrasse, enjoying card games and even horseplay in the lawn, well into his advanced years.[11]

And here, in 1905, he reviewed and decided to publish the miraculous and fog-cutting voice of a young outsider, Albert Einstein, a man with neither a PhD nor a university position. At the same desk, Planck enthusiastically penned his first lectures on Einstein's groundbreaking Invariance Theory—or as Max called it, "Relative Theory." At the scientifically ripe age of 47, when many minds no longer bend, Max embraced Einstein's radical new notions of motion, in which rulers changed their length and events changed their order depending on the observer's own motion. It is no exaggeration to say that Planck discovered Einstein, and he helped shepherd the squarest peg of Einstein's genius into the round hole of the science establishment. Max stood with young Einstein and said they had to stick together against waves of skeptics. In time, Planck persuaded Einstein to join his physics powerhouse in Berlin.

In 1905, Wangenheimstrasse 21 nurtured the next generation of Plancks. The worrisome Karl was approaching the end of his secondary studies and thinking of pursuing geography in college—surely, his parents thought, that would finally provide some direction for him. The identical twins Grete and Emma were now 16 and prone to soothing the home with their singing and the expert bowing of their violins. Whether or not he could always distinguish their voices from one another, Max could probably hear their mother's songbird timbre in both. And finally there was 12-year-old Erwin, his father's favorite. The young politician already mediated the thorny dialogue between his father and older brother, Karl. Their father had just given Erwin a cello and fully welcomed him into the family's music.

Max and Marie hosted parties for a mix of professors and musicians. Max, an exacting and practiced pianist, often performed with other guests in duets, trios, or quartets. He loved especially the work of Schumann and Schubert, German romantics, and he counted the renowned violinist

Joseph Joachim as a close friend. In time, the Planck home featured trios of father Max, son Erwin, and their friend Einstein on violin.

In October of 1944, such memories must have floated like surreal phantoms to Max Planck. Allied bombing fury obliterated all traces of Wangenheimstrasse 21 and the city around it. German science lay in similar ruins, as the Nazis had chased scores of the most talented scientists away and redirected most of the remaining funding to building war machines. Planck had either resigned or had the Reich usher him from his remaining scientific posts, while attacks on his character, and even his scientific legacy, continued. Nazi sympathetic scientists now accused Planck of exaggerating his own importance and stealing credit from more deserving scientists. They said Planck claimed victories for his puffy calculations that rightly belonged to sweat-filled laboratories. Even his Nobel status provided no protection—Hitler had proclaimed the Nobel prizes unworthy and forbade Germans from accepting them. Despite Planck's tension with the Reich, his allegiance to Germany had cost him a dear friendship with Einstein—their once warm dialogue had long fallen silent by 1944. Of all the musical partners from Wangenheimstrasse, only one other remained: Erwin.

Now Max Planck's closest friend and the last remnant of those best days awaited death by hanging. Within two days of the sentence, Planck had also written letters to Heinrich Himmler, the powerful leader of the Nazi's SS, the *Schutzstaffel.* Himmler, who ran the concentration camps, could show mercy at unexpected times, especially for familial appeals or connections.[12] Against all odds and logic, the great physicist gathered the shards of his optimism and hoped for one madman or another to make a charitable gesture.

2

April 1943

Planck had last written to Hitler 18 months earlier, long before the briefcase bomb. Erwin was a free man then, or as free as anyone could hope to be in wartime Nazi Germany, and Max Planck's chief concern was surviving the onslaught of bombing. Newly a refugee from Berlin, Planck sat with stacks of birthday greetings on April 23, his 85th. He took several days and answered each card and telegram by hand, including that of the Führer.

Allied bombing had cracked open the Planck home in the early spring of 1943, damaging the roof. With holes providing new and unwelcome views of Berlin's winter sky, Max and Marga, his sturdy second wife, fled Wangenheimstrasse 21 in favor of Rogätz, a small village some 80 miles southwest of Berlin.[1] Max abandoned his cherished study and library. He evacuated nearly 40 years of memories in the home: duets with Einstein, a son returning wounded in the Great War, and many a Christmas watching candle flames wink out one by one on the tree. He left behind photos of his first wife and the garden where Max played tag with his children and guests well into middle age. As he'd always done much of his work at home, he exited the command center from which he'd fought to build and maintain German physics. Max and Marga left most of their belongings, no doubt assuming they could return after ordering some repairs. But if they had stood flanked by suitcases, facing the house, they would have seen the last of the villa, a sad and damaged version.

In Rogätz, Max felt his age. Though the most robust of his siblings by far (he outlived them by many years), and a dedicated alpine hiker into his eighties, the grip of age pressed him now, curling his back. It jabbed him without mercy during his treasured daily walks. But Planck, as usual, emphasized the positive, writing of exile as an unexpected blessing with an expanse of free time. And although he missed seeing Wilhelm Furtwängler conduct the Berlin Philharmonic, he learned to enjoy the same concerts via radio.[2]

He looked forward to visits from Erwin and Erwin's wife Nelly. He would write to them in Berlin, describing the plump strawberries or cherries awaiting them in Rogätz. Planck was particularly fond of his daughter-in-law. After completing her medical training later that year, she took the examinations to become a doctor, acing everything but a special Nazi component on recognizing racial characteristics; while this must have frustrated her at the time, it cheered her father-in-law.[3] He had come to cherish Nelly as he would his own daughter, and he was happy to see none of Hitler's obsessions infecting her mind.

As he answered Hitler's telegram that April, he would have struggled for any positive small talk. For the last 10 years, the Nazis had beaten his optimism into submission. He had initially appealed to his colleagues to be patient, to wait out the radical tendencies of any new government, only to have this new government prove him foolish at every turn. In 1943, did he think of his first thank-you card to Hitler? After receiving birthday greetings from the new Reich's Chancellor in 1933, Max had used his thank-you note to request a meeting. He aimed to politely make his case—he had some stellar Jewish colleagues after all—and the National Socialist German Workers' Party could not be as monstrous as Einstein and the leftists claimed. But 45 minutes alone with Hitler had shaken the normally imperturbable Planck to his roots.

Now he was simply one of millions of refugees spilling from the Führer's madness, joining his Jewish friends and fellow physicists Albert Einstein and Lise Meitner in flight, albeit more comfortably. In Rogätz, Max and Marga were honored guests of steel magnate Carl Still. The guesthouse, a massive, white manor only steps from the Elbe River, stands in Rogätz today. Carl Still built his vacation manor next to a thick stone tower, the last remnant of an eleventh-century castle. Together the structures now serve as the town's primary tourist attraction.

Planck referred to the entire guesthouse as their "magic castle" and greatly enjoyed being pampered there, but Marga fretted about their abandoned home in Grunewald.[4] River view or no, the deluxe lodgings couldn't keep her from missing their life and home in the city. She had relished the hosting and entertaining that came with a husband owning Planck's superlative stature and network.[5]

From what we know of Planck, he would have missed his routine and his library more than the old house itself. Time, family, and work were top priorities, but comfort was never a consideration. As a commuting professor, he'd always taken a third-class train ticket. Former student Gabriele Rabel, bolder than her classmates, used to walk with Planck to the train station after his lectures. She recalled her shock at seeing him enter the third-class compartment. "No cushions," she said. "Nothing but unmitigated wood."[6] Certainly, his lifelong preference for the simple and unadorned helped Planck adjust to life as a permanent guest, but his earliest years would have made the smaller town familiar as well.

At the time, Rogätz could remind one of Eldena, another rural town further along the Elbe River's meander to the North Sea. When Max was a boy, his parents made Eldena the family's vacation nest, and he recalled summers there with great fondness into his later years. Playing croquet with his siblings and cousins, reading literature aloud, and putting on their own musical plays consumed the long, warm days.[7] In darker times, nearly 50 years later, he wrote to a cousin about these, "days and weeks that not only stay in our thoughts but grow in glory … as we become conscious of the incredible good luck that surrounded us, and how deeply grateful we must be for what we took for granted then."[8] Though one could hardly call these times the peak of Planck's extraordinary arc, his childhood provided a period of relative comfort and calm. His years with his parents and siblings belied the later tone of his life.

He was born Max Karl Ernst Ludwig Planck in 1858, joining a large, pragmatic family. (Apparently, church records in Kiel show him baptized as Karl Ernst Ludwig *Marx* Planck, perhaps via clerical error.)[9] Cooperative, communicative, and well liked, he soon answered to just "Max." He enjoyed a privileged childhood, with nannies, a cook, and housemaids attending to the family. The home at Küterstrasse 17 buzzed with servants and children. His father, Wilhelm Planck, lectured as a legal scholar at the University of Kiel, and his mother Emma bore five children to Wilhelm. Max was the fourth of Emma's children and the sixth of seven children in the house, as Wilhelm had two from his previous marriage. When Max arrived, his full sibling Hermann was seven, Adalbert five, and Hildegard three (Figure 2.1). His older half-siblings Emma and Hugo were starting their teen years. Little Otto would follow five years later.

Max was particularly fond of his outgoing mother Emma; they maintained close contact until she died at age 93.[10] Of all his siblings, he was apparently attached in particular to Adalbert, his older brother by several years and an engineer by trade. He and Adalbert eventually scheduled their

Figure 2.1. Four children of Wilhelm and Emma Planck. From left to right, Adalbert, Max, Hildegard, and Hermann.
Courtesy Archiv der Max-Planck-Gesellschaft, Berlin-Dahlem.

weddings to be just days apart in Munich,[11] and into the last years of his life, Max exchanged fond letters with Adalbert's children.

Kiel was a small fjord-bound town on what is now Germany's northern coast, but the nation didn't exist at the time. Rather, the German Confederation loosely bound 39 separate German-speaking states with a common constitution. Most notably, the Confederation included the Austrian Empire and the Kingdom of Prussia, but it also linked Saxony, Bavaria, Hesse, Baden, and so forth. Kiel sat within the Schleswig-Holstein region, owned then by the King of Denmark, and the town housed just 15,000 souls, many of them involved in the shipping trade.

We have little information of Planck's life in Kiel, but we can understand his earliest environs via Thomas Mann's first novel, *Buddenbrooks*. Mann set his semi-autobiography in Lübeck, a similar-sized town just 50 miles down the jagged Baltic coast from Kiel. The novel's long span includes the period of Planck's childhood, and its characters move within the same comfortable strata of Prussian society. Horse hoofs knocked and scuffed methodically against cobblestones, and busy days raised dust thick enough to obscure any view of the ocean. The novel's main character and her friends, idle youths all, "would clamber about in the granaries on the water-side, among the piles of oats and wheat, prattling to the labourers and the clerks in the dark little ground-floor offices; they would even help haul up the sacks of grain." The child, "knew the butchers with their trays and aprons, . . . accosted the dairy women when they came in from the country, and made them take her a little way in their carts. She knew the grey-bearded craftsmen who sat in the narrow goldsmiths' shops built into the arcades in the market square; and she knew the fish-wives, the fruit- and vegetable-women, and the porters that stood on the street corners chewing their tobacco."[12] On a spring or summer day, women like Emma Planck donned white stockings and strap shoes. They shielded themselves with straw hats and parasols bordered in lace. Men of Wilhelm Planck's class appeared in well-cut blue suits, standing and smoking with their hats rocked to the backs of their heads. At home, families would celebrate spring in the garden after the regular afternoon coffee and cake, while children might, if they were good, receive "the bishop," a red punch. After dinner, women worked embroidery by candelabra, and young boys like Max Planck would earnestly absorb old cultural maxims such as, "attend with zeal to thy business by day; but do none that hinders thee from thy sleep at night."[13]

Presaging a life repeatedly scarred by war, the child Max Planck witnessed soldiers marching through Kiel.[14] In 1864, Austrian and Prussian troops drove the Danes from the region, and Kiel then belonged to King Wilhelm I of Prussia. The town began to grow as Wilhelm decided to park his Baltic fleet there. One can imagine a late March day in 1865 when the fleet first sailed into harbor, most probably in drizzle or even snow. Many townspeople would have gathered to watch and welcome, but the fleet was only 11 vessels at the time, primarily powered by sail. The local firm Maschinenbauanstalt Schweffel & Howaldt had already revamped its steam engine production to suit ship builders, and Kiel went on to major naval prominence. Germany tested the first of its famous U-boat submarines in Kiel's fjord, and in 1918, the German naval mutiny started there, precipitating the end of World War I.

Further conflict hit Germany in 1866. The German Confederation fractured, as Prussia and its northern allies made war against Austria and the southern states. Prussia prevailed, and while it held itself from overrunning its southern adversaries, it formed a new confederation that excluded

Austria, Bavaria, and their neighbors. The Plancks soon moved into the banished territories. In 1867, soon after Max turned nine, his father accepted a post at Munich's prestigious Ludwig Maximilians University and moved the family south, from Prussia to the Kingdom of Bavaria.

The young Max showed aptitude for everything he tried, with a fondness for languages, mathematics, Bible study, and music. His teachers called him a "favorite" of staff and students alike, and he won multiple catechism awards for his religious studies.[15] He had a high voice of excellent pitch, and he possessed great skill at the pianoforte, as the piano was known at the time. He played organ for his family's Lutheran church and sang soprano for both his school and church choirs.

Daily practice was surely required (and apparently relished), by young Max. In the era of Max's training, attitudes were typified by the renowned Austrian pianist Emil von Sauer. Discipline, concentration, and finger strength were the primary goals for any student. Sauer said that a concert pianist's "fingers must be as strong as steel, and yet they must be as elastic and as supple as willow wands." He prescribed repetitive doses of the "Pischna Exercises," a progressive set of 48 to 60 technical pieces that required advancing level of finger flexibility.[16] For young Max, a professional career appeared possible, even likely. Planck enjoyed composing as well. Even in college, when he saw no turning back from physics, he wrote an operetta called *Die Liebe im Walde*, or Love in the Forest.

His reported struggle to write original melodies helped Max see that he had little future as a composer, but music stayed with him throughout his life. He maintained an active reputation as a concert-grade pianist, often performing at social functions. And when he finally fell out of practice, his ear was so well trained that listening to some performances was painful, given a swarm of tiny mistakes. He reportedly delighted in losing this level of precise hearing in his last years, when he could then embrace recorded music more openly.[17]

In the summer of 1870, France declared war on Prussia, and the cunning German Chancellor Otto von Bismarck correctly saw his long-awaited opportunity to unify all the German-speaking kingdoms and duchies into one empire. Prussia and her allies swept to victory within six months with relatively light casualties. Germans could jointly relish returning Napoleon's old favors, but the Planck household mourned. Presumably as part of the Bavarian infantry, Max's older brother Hermann died near Orleans following a fierce early December battle.[18] By the end of January, the war was over, and Prussia's king became Emperor Wilhelm I of a newly united Germany. The Planck family would see its own blood in the stitches of this new empire, and like so many others, they were intensely patriotic for life.

In Munich, Max enrolled in the Maximilians Gymnasium (a secondary school, aiming to the endpoint of an *Abitur*, a mandatory entrance exam for university study). Though teachers praised his social skills, potential, and penchant for logic, he was never at the top of his class. In the yearly rankings, Max floated between the top third and the top fifth.[19] When he passed his *Abitur* examinations at age 16, however, he earned one of the top scores in his class, and he was ready for university lectures in Munich.

While at the gymnasium, Planck came under the spell of Professor Müller, who primarily taught mathematics with a steady sprinkling of physics. Even late in life, Planck recalled Müller with a sort of reverence. "I shall never forget the graphic story Müller told us, at his *raconteur's* best, of the bricklayer lifting with great effort a heavy block of stone to the roof of a house." Every bit of energy expended by the bricklayer remained in the block, for years, even centuries, and was released when the stone eventually plummeted back to Earth—this was conservation of energy brought to life. It was an absolute law of nature, independent of the bricklayer's identity or the building's structural choices. Only the weight and height of the stone mattered. This truth was "absorbed avidly, like a revelation," by the eager mind of Planck, and he was forever captivated.[20] Soon he would relearn this principle as the first law of thermodynamics, a new field that recruited Planck as its devoted disciple, shaping the rest of his life's work.

Spending his first three years of university in Munich, he studied mathematics and laboratory physics, since, "there were no ... classes in theoretical physics as yet."[21] He took much of his instruction from the physicist Philipp von Jolly, a man who also delivered famously ironic advice. As Max expressed the will to pursue physics as a career, Professor von Jolly cautioned him against such a dead end. He said that physics was nearing its natural conclusion, given the success of Isaac Newton's mechanics and James Clerk Maxwell's work in electricity and magnetism, topped by the new and nearly complete field of thermodynamics. There was simply nothing left to do. But Planck would not be dissuaded, and he told von Jolly he didn't mind if his life's work would be largely an end in itself, for further and deeper edification. He would happily tend the garden of physics.[22]

Although Planck would soon learn just how wrongheaded a professor's advice could be, he would also learn with disappointment that an academic career would need more than hard work and concentration. And while he would grow to see von Jolly and his other Munich professors as limited and provincial, he would also find the doors of the upper echelon closed to him for years to come. From the establishment, Planck found "no interest, let alone approval" for most of his earliest work.[23] And late in life he admitted that his career's first works had been done better, and earlier, by an American physicist named Willard Gibbs (not widely read in nineteenth-century Germany). Overall, compared with other revolutionary figures in physics, Planck was a late bloomer. Although the Newtons and Einsteins of history made their marks in their twenties, Max Planck, as we shall see, spent many of these years living with his parents, wondering if he could ever win a paid position.

Perhaps these early struggles nudged him a generation later to be an encouraging voice to young scientists brimming with new ideas. In his middle age, he advocated for the peculiar yet eloquent work of the young Albert Einstein, and he stood against the establishment to find a scientific place for Lise Meitner, the brilliant young Jewish woman who came to Berlin with no plan beyond hearing Planck speak.

And in 1943, sifting through hundreds of birthday greetings, he might have paused to double-check for a card from Stockholm, Meitner's refuge after escaping Nazi Germany. He missed his friend Lise, but he would see her soon enough.

3

Late May 1943

In the early summer, Planck took a rare wartime trip outside of Germany, traveling to Helsinki and Stockholm. Presumably, he went by train, tracking a similar route as he had to receive his Nobel Prize. Balancing on elevated tracks, he crossed a watery maze through Denmark, clacking along the border between the North and Baltic Seas. The train stopped in Copenhagen and then continued northward across lake-flecked lower Sweden.

Planck had confided in a colleague that by this time he was too ashamed of his nation's conduct to attend regular scientific conferences. But in this case, he could renew some old conversations in smaller settings. Headlining his list was the physicist Lise Meitner, a dear friend of many years. Max had not seen Lise since her sudden and night-covered escape from Germany five years earlier. As a Jew, she had pushed her luck for many Nazi years in Berlin, out of devotion to her scientific work and colleagues there.

Even in 1920, on the way to collect his prize, Max Planck could observe the seeds of the madness to come. With a heavy personal heart, following a string of tragedies spanning World War I, Planck went to claim a Nobel Prize in Physics that most considered long overdue. He and Marga had traveled with the eminent chemist Fritz Haber and his second wife, Charlotte, as Haber had won the Nobel Prize in Chemistry. But the German contingent did not end there.

Johannes Stark, a rash and increasingly political physicist, went to collect a Nobel Prize, but unlike Planck, he won for his laboratory measurements.[1] Although Stark had once been a Planck supporter and one of the earliest to embrace Einstein's revolutionary ideas, he had dramatically changed his tune by 1920. He resented Planck's funding of theoretical work, for one, but more disturbingly, he gave a scientific voice to growing anti-Semitism. Months after the prize ceremony, he decried Planck's circle as a Jewish clique obsessed with mathematics. The pejorative verbal linkage of math-heavy physics with Jewish physicists would only grow within Germany. Stark and another Nobel-winning physicist, Philipp Lenard, were busy advocating the "Deutsche Physics" movement, meant to purify German science by rinsing away misleading and overly mathematical Jewish influences. An early resonance emerged between scientists like Stark and a young firebrand speaker, Adolf Hitler, who had just joined Anton Drexler's fringe political party, the NSDAP (*Nationalsozialistische Deutsche Arbeiterpartei*, or the National Socialist German Workers Party). Hitler was a veteran of the trenches from World War I, convinced that a Jewish conspiracy had led to Germany's defeat. He plucked some of the scientific resentment from Stark and others, molding it for his own rhetoric. "Science, once our greatest pride," Hitler wrote in a 1921 newspaper screed, "is today being taught by Hebrews."[2] Many mocked Hitler in the wake of his 1923 failed coup, the "Beer Hall Putsch," referring to him as the "little corporal." But by 1924, Stark and Lenard would be among the first Germans to pledge fervent support to the radical politician.[3] Stark's rabid alliance with the Nazis would come to menace Planck in time.

A silent film shows all of these scientists assembled for the June 1920 Nobel ceremonies in Stockholm. Haber and Stark stand at nearly opposite ends of the frame, and the tension may have already pulsed silently at the banquet. Haber, a brilliant chemist with a career both wondrous and horrifying, was a Jew. He had also literally transformed humanity and the planet's population when he learned to extract nitrogen from the air, revolutionizing our ability to fertilize crops. But he later turned that same genius to developing chlorine gas weapons for clearing enemy trenches in World War I. And on the first day of its successful use against the Entente troops, his first wife took up his pistol. Standing in front of him, she ended her life. But as a committed patriot living in the greatest war mankind had yet seen, he soon returned to his work. Haber was long focused on assimilating, but only his heritage would have mattered to Stark in 1920. And in the end, that's all that would matter to Haber's lifelong home, Germany.

In 1943, the flat and watery expanse of lower Sweden blurred past Planck's train window. Much had changed since 1920. Haber had fled Nazi Germany and found a sad death in exile, while Stark had boldly attacked Planck's science (overrated), and his heritage (a "white Jew," and perhaps a real one). The Nazis had also separated Planck from one of his favorite colleagues, Lise Meitner. Now he would have the happy chance to see her again.

According to Meitner, they enjoyed a wonderful day together. She reported to a friend that Planck's lecture in Stockholm, "left a strong impression because of its clarity and due to the size of his personality."[4] She wrote to a mutual friend that, "It was great meeting him of course. Great pleasure and a joy. But he has also aged." It was hard to see him without his renowned, nonstop energies intact. But "he got lively here and there. And then I was so overjoyed to feel the old Max Planck."[5]

Meitner, an Austrian Jew, perched in Stockholm as an unhappy immigrant, missing her Berlin laboratory, her scientific colleagues, her evenings at the Planck home, and long conversations with Max. Her own story has spawned at least two excellent biographies, but she provides a rare window into the life of Planck as well.[6]

Born in the Vienna of 1878, Meitner, like Planck, came from a large and musical family that celebrated learning. Unlike Planck, she wanted a life in science from her earliest years and had to swim upstream to find it. As a young girl, she would hide mathematics books under her pillow to read long after bedtime. And her sisters would sometimes tease her: "Lise, you have just walked across the living room without studying!" The education of girls in Vienna stopped cold at age 14, but Lise kept to her books and awaited the glacial thawing of these old boundaries. In 1901, she was allowed to apply to the University of Vienna, as long as she crammed the equivalent of an entire eight-year "gymnasium" education into two years. She did this, flew through college absorbing 25 hours of science lectures per week, and in 1906, became the first woman awarded a physics doctorate from the university.[7]

Along the way, she came to cherish one of her professors there, the mercurial and passionate Ludwig Boltzmann, a bowling ball of mathematical physics talent. His wide, animated face, rimmed by curly black hair and a reddish beard enraptured his students. He pledged to give his students "everything I have," in return for their devotion to the material. His lectures were "the most beautiful and stimulating that I have ever heard," Lise later recalled. She "left every lecture with the feeling that a completely new and wonderful world had been revealed."[8] And indeed, for physics he was ahead of his time; as an ardent believer in atoms and a man convinced of taking statistical approaches to complex problems, he tried to reveal a new way forward. Boltzmann felt his energy and innovative thinking bottled by walls of more standard physics belief around him. With doubt raining on him in Vienna, he roiled in frustration and feelings of persecution; by all evidence, he simultaneously wrestled with the demons of depression.

Boltzmann is a crucial figure in the story of Max Planck, even though they were in no way close friends or friendly colleagues. Planck was only offered the coveted post in Berlin after Boltzmann had turned down the university's offer.[9] Most importantly, the mathematical tools Boltzmann employed for his efforts to describe containers full of bouncing energetic atoms later came to inspire Planck in his groundbreaking discoveries. Planck knew Boltzmann's work well and made many analogies to the Austrian's efforts. Despite a couple of notable and public tiffs, in the end they respected one another's views a great deal. But Boltzmann did not just lay the mathematical groundwork for Planck's own work and what became quantum mechanics—he indirectly introduced Planck to Meitner.

When depression finally overtook Boltzmann in 1906, he hanged himself during a family vacation to Italy. Sadly, the 62 year old did this precisely at the dawn of physicists broadly accepting the existence of atoms. The University of Vienna immediately tried to recruit the only physicist who could properly fill the position: Max Planck. Lise Meitner attended Planck's invited lecture in Vienna and when he decided to remain in Berlin, she was intrigued enough to seek him out and hear more. Besides, Boltzmann had always spoken highly of Berlin, if not Planck himself. At first, Meitner intended just a year away from her dear family, to see what she could learn from Planck.

In 1907, she arrived at the University of Berlin without much of a plan, and she discovered that women were typically barred from Prussian universities at the time. She met with Planck and asked to attend his lectures; he agreed without hesitation. Although that might paint him in a progressive hue, his true thoughts were not so far beyond mainstream male physicists of the time. Responding to a survey about academic women in 1897, he had said he would allow the occasional talented and driven woman to attend physics lectures. However, he also said the following.

> I must hold fast to the idea that such a case must always be considered an exception, and in particular that it would be a great mistake to establish special institutions to induce women into academic study, at least not into pure scientific research ... in general it can not be emphasized strongly enough that Nature itself has designated for woman her vocation as mother and housewife, and that under no circumstances can natural laws be ignored without grave damage.[10]

And privately he maintained such traditional views for years to come. In a 1909 personal note to some of his college friends, he wrote of a colleague's home where, "the daughters were studying, and you talk with them as with young men. I'm not very sympathetic to this. But I mean, you should let everyone pursue their own bliss."[11]

Nonetheless, Planck would tirelessly advocate for Meitner over the next 30 years. He would find a way to wedge her into an otherwise closed, wholly male scientific community, create her first paid position, and even champion her becoming the first female science professor in the university's history.

Lise Meitner later wrote that, "In his outward behavior, Planck was very reserved, for all the affection he inspired. Some people mistakenly regarded this as a sign of conceit, but nothing could have been further from his character. He had a rare honesty of mind and an almost naïve straightforwardness."[12] She and others who got to know the man beyond first impressions spoke of him with "near reverence." They mentioned the transformation of Planck in more informal settings. He "was such a wonderful person, that when he entered a room, the air in the room got better. . . . The younger generation of Berlin physicists . . . felt this very strongly."[13]

Young Lise the student also realized that Planck was no Boltzmann. At first, she was "a little disappointed in Planck's lectures," and though she praised their clarity and construction, "they sometimes gave a rather colorless feeling."[14] Where Boltzmann had been funny and animated, sprinkling lectures with personal anecdotes, Planck seemed completely controlled, creating perfect but, by comparison, sedate lectures. His first impression on students was often one of an aloof or arrogant thinker. Personal anecdotes and humor were to be strictly separated from the physics classroom.

Meitner's parents sent their student daughter an allowance, and she earned additional money as a book translator and science writer for nontechnical journals. She rented single rooms and spent as little as possible, except for newspapers, cigarettes, and nosebleed seats at the symphony.[15] This modest living continued for five years as she remained in Berlin, drawn increasingly into a new type of research: radioactivity. Here, her physics talent could push forward the work of chemists.

At a weekly physics colloquium, Lise met another young scientist named Otto Hahn. Like Meitner, Hahn had just arrived in Berlin. He had already discovered several new radioactive elements while studying in laboratories in England and Canada. The two began talking excitedly about further experiments with radioactivity. What other materials might be radioactive? How can we measure the emitted radiation itself? Can we put this mysterious energy to use? They needed Hahn's chemistry expertise to precisely identify the materials and measure their exact amounts, while they needed Meitner's physics to diagnose the radiation products themselves and attempt a physical model of the underlying processes. In essence, Hahn could determine the make and model of a material's car, while Lise would look at what came from the tailpipe in hopes of figuring out how the engine worked.

At the time, radioactivity was a new and nearly magical topic, having been discovered only a decade earlier. By the time Meitner and Hahn joined forces, pioneer Marie Curie had won her first Nobel Prize (Physics, in 1903, shared with her husband Pierre). In the earliest days, the Curies and others would study radium in every conceivable way, including holding samples against their arms and recording details of the resulting burns. They were giddy with wonder.

Planck helped Meitner pursue her bliss. He appealed to his colleague Emil Fischer, director of the university's Chemistry Institute, to bend the strict rules forbidding women. Fischer needed persuading, because he saw specific laboratory dangers for women. Meitner recalled that, "his reluctance . . . stemmed from his constant worry about a Russian student whose rather exotic hairstyle might catch fire on the Bunsen burner! He finally agreed to my working with Hahn, if I promised not to go into the Chemistry Department where the male students worked."[16] Meitner could then enter the chemistry building basement, heading for the old woodshop where she could pursue research with Otto Hahn. And since the chemistry building had no bathroom for women, she trekked periodically to a restaurant down the street. But she had found her calling in their collaboration (Figure 3.1).

Studies of radioactivity, "were then developing incredibly quickly; nearly every month brought a wonderful surprising new result from one of the laboratories," she later wrote. And she struck up a rare and playful camaraderie with Hahn.

> When our work was going well we sang duets, mostly Brahms Lieder, which I could only hum, while Hahn had a very good singing voice. . . . If he was in an especially good mood he would whistle large sections of the Beethoven violin concerto, sometimes purposely changing the rhythm of the last movement just so he could laugh at my protests . . . we had a very good relationship

Figure 3.1. Otto Hahn and Lise Meitner in their Berlin laboratory, circa 1920.
Photograph from *Schoepfer des neuen Weltbildes: grosse Physiker unserer Zeit*, by Hans Hartmann, Bonn: Athenaum-Verlag, 1952. Courtesy AIP Emilio Segre Archives, Brittle Books Collection.

> with the young colleagues in the nearby physics institute. They often came to visit us, and sometimes they would climb in through the window of the wood shop.... In short, we were young, contented and carefree, perhaps politically too carefree.[17]

She and Hahn worked as equals for the next 30 years, combining their separate expertise to discern what exactly was happening inside materials that sent out stinging, ballistic energies. After the death of her father, when her allowance stopped, Max made her his physics assistant in 1912, giving her a paid position and crowning her as the first female academic assistant in the university's history. In time, with Planck championing her cause, she became a Privatdozent in physics in 1922, allowing her to formally teach classes. As a standard part of the promotion, Lise gave a public lecture. Scientists had begun to recognize that some of the radiation arriving from outer space was identical to what flew out of radioactive atoms in the laboratory, and she gave a talk on the "Significance of Radioactivity in

Cosmic Processes." But the local newspaper reported her talk on "*Cosmetic* processes," providing one of Meitner's favorite woman-in-science stories.[18]

In much shorter order, she'd become a fixture within the Planck family, having met them soon after her arrival in Berlin.

> I was invited to spend an evening at his home in the Wangenheimstrasse, where I met his first wife and his twin daughters . . . who were to become my close friends. [Max Planck] was a very pleasant man; he liked to have people around his home, even when I was attending his courses. And as soon as we were in his home he played tag at least as eagerly as we young students: he tried to catch us while we ran, really he did. He could run very well—he had long legs—he was really a very agreeable man. He wasn't arrogant . . . a Prussian, the way one imagined him—not at all.[19]

The vision of a playful Planck is still difficult to fit with either his reputation or the sour countenance we find in photographs. By all accounts, he had a terminal case of probity and a nearly doleful seriousness in public life. But Meitner's memories paint the man in his home and family. Ten years the twins' elder, Lise bonded with them, scheduling regular picnics, and eventually hiking along with the Planck family's mountain vacations.

Meitner and Planck were certainly more sober when they met in 1943. Meitner had fled Berlin after the German annexation of Austria in 1938 (known as the *Anschluss*, meaning "the joining"). This move shredded the protection provided by her Austrian citizenship. Despite having converted publicly to Protestantism in her early Berlin days, the Anschluss left her with no refuge from the worst Nazi policies. A small network of colleagues had helped get Lise across the Dutch border in secrecy, just as the Reich learned that she might flee. "So as not to arouse suspicion," she later wrote. "I spent the last day of my life in Germany in the institute until 8 o'clock at night, correcting a paper to be published by a young associate. Then I had exactly 1 1/2 hours to pack necessary things in two small suitcases."[20] Hahn helped her pack and took her to his home for the night, in case anyone looked for her in her apartment. They told no one of their plans, even Planck. Hahn had informed everyone at their institute that she was headed to see her family; he even marked his calendar "Lise goes to Vienna." That night, they came up with a simple code by which she could telegram success or failure at the border crossing. Hahn recalled, "People trying to leave Germany were always being arrested on the train and sent back." As her trip approached, "we were shaking with fear that she would not get across."[21] At the last moment, Hahn gave her his mother's wedding ring, so that she might have access to emergency cash. Though she nearly turned back at the border, her crossing on July 13 went smoothly, and only two weeks later Hahn was being asked about his missing Jewish colleague. Nazi officials intercepted her crated furniture and effects, and when she received her belongings 10 months later, she found the furniture in splinters, dishes in pieces, and ripped pages of books tossed throughout the mess. She summarized her Swedish purgatory in a letter to a friend. "One dare not look back. One cannot look forward."[22]

What did Planck and Meitner talk about in Stockholm? We have only an enigmatic reference from Lise's letters. In privacy, Planck confided that they were of like mind—he knew that Germany must be defeated in the war. "We have done the most horrible things," he said to her. "Terrible things must happen to us."[23] Planck was not one to make bold pronouncements, but he was correct on both counts here.

What did he know of "horrible" deeds? He'd seen his colleagues, including Lise, driven from the Fatherland. He'd seen the ever-advancing thresher of new laws, first limiting, then persecuting, then gathering Jews along with other Nazi targets. He must have known of the inhuman trains, even if he didn't know their destination. Starting in 1941, 50,000 Berlin Jews left via the Grunewald station, many in windowless cattle cars.[24] This was but two kilometers from the Planck home on Wangenheimstrasse, and Platform 17 there now presents a memorial to the grim departures. If nothing else, Planck knew the forced emigration of scientists too well. And he was smart enough to see past Reich rhetoric and the fake pretenses used to ignite war. But new evidence suggests he would have realized a sickening amount more. In 2013, research published by the Holocaust Memorial Museum documents a much greater proliferation of Nazi camps, subcamps, and enforced ghettos: some 42,500 across Europe. "You literally could not go anywhere in Germany without running into forced labor camps, P.O.W. camps, concentration camps," Dr. Martin Dean has told the *New York Times*. "They were everywhere." The group's research documents 3,000 camps, mini-camps, and "Jew houses" in the Berlin area alone.[25] And since Planck saw bombing damage, both in Berlin and Kassel, he would have noted the camp victims forced to clear rubble.

Immediately after the war, Meitner exchanged letters with her old friend and colleague Otto Hahn, and they fell into arguments about Germany. Meitner would not accept his claims that he didn't know about the Nazi atrocities. "You had told me yourself in March 1938 that (a colleague) had said to you that terrible things were being done to the Jews. So, he knew about all the crimes that were being planned and later executed and was a member of the Party despite it; and you regarded him as a very decent person." She didn't spare her old friend, reminding him that he "never had any sleepless nights."[26]

But could she have said something similarly reproachful to Planck? Her reaction to Planck's "We have done the most horrible things" statement suggests not. "He used the words 'we' and 'us,'" she wrote to a friend, noting his inclusion of all Germans, as opposed to just Nazis. "And yet this 85-year-old man was more courageous in his resistance than all the others."[27] For many, it will be hard to accept the lean old physicist as courageous.

Flanked by swastikas, he praised his Führer to start speeches, executed the Nazi salute, and when asked to bar Jewish students from his classrooms and then fire Jewish staff, he did as he was told. In 1937, Lotte Warburg attended one such speech in Switzerland. As the daughter of deceased physicist Emil Warburg and brother to physicist Otto Warburg, she felt free to judge what she saw. After he avoided the mention of any Jewish physicist (including her father) and ended his talk by thanking Hitler, she screamed to her journal, "Such a monstrous lie! … everyone will say … 'Ah Planck! there's an honorable character!' … no one will get at the truth, at the monumental cowardice and spinelessness with which his last years were filled."[28] Planck was 75 when Hitler came to power. He stood with his Fatherland, no matter its warts and crimes, even at the cost of friendships. In summary, Planck dedicated himself to the idea that he could do the most good from within the Reich. His dedication to this notion paralleled his dedication to a homeland that would endure and overcome anything, be it war, famine, or Nazi leadership.

Though she was critical of most German scientists after the war, including friends like Otto Hahn, Meitner maintained a different assessment of Planck until her last days. "He had an unusually pure disposition and inner rectitude," she wrote in recollection, "which corresponded to his outer simplicity and lack of pretension."[29]

Beyond what they discussed of the war, it would be hard to imagine this June 1943 meeting without some science talk. In the preceding years, Meitner had played a leading role in diagnosing, though arguably not *discovering*, nuclear fission, the splitting of the once unassailable atom. In fact, it was her conversation with nephew Otto Frisch that took the term fission from biology (the splitting of one cell into two) and carried it to the innards of the atom, which could divide in a similar way when under extreme duress. Despite our standard textbook pictures, physical evidence suggests that the nucleus is more like a hovering water droplet than a clump of plastic particle grapes.

In 1938, shortly after her escape, Otto Hahn sent Meitner some puzzling new results and asked for her intuition and input. (One can imagine their frustration at the separation after 30 years of trading ideas in person.) After bombarding small bits of uranium with neutrons (the neutral versions of protons), Hahn and a colleague found apparently new elements that they could not identify. It was thought that adding neutrons necessarily swelled a nucleus, an act of force-feeding, so that all the experimental products had to be "trans-uranic," or heavier than uranium. Hahn was very confused by one leftover material from the experiments: It behaved just like an alkaline earth metal such as calcium or magnesium, but what could it be? Though it may strike us today as obvious, no one had seriously considered the idea that the nucleus could *split* into two smaller chunks, with necessarily different characteristics.

Frisch, a physicist, recalled the critical conversation with his aunt. After looking over Hahn's data and scratching their heads, they were taking a break for fresh air. "We walked up and down in the snow, I on skis and she on foot . . . and gradually the idea took shape that this was no chipping or cracking of the nucleus but rather a process to be explained by Bohr's idea that the nucleus was like a liquid drop; such a drop might elongate and divide itself."[30] Frisch came up with the term fission, and back indoors, Meitner went to work on the energies involved. She was the first to realize what the experiments implied: an enormous—even ridiculous, or disturbing—source of power. They communicated their thoughts to Hahn but did not publish with him. Hahn and a colleague in Berlin presented a new paper with the experimental results, and, after some delays at a scientific journal, Frisch and Meitner published their more theoretical ideas, including comments on the enormous energies released. This separation—now both intellectual and geographic—would have significant consequences in terms of recognition.

Even in 1943, Planck and Meitner may have reasonably discussed the good chance of her sharing a Nobel Prize with Otto Hahn for her part in the fission discovery and to acknowledge decades of fruitful collaboration. Planck had nominated them both, multiple times.[31]

So too, they could have discussed the future uses of nuclear fission. Meitner knew the Americans were working on a bomb. She had just visited Chicago in 1942 to attend what she called "a caricature of a scientific meeting," with everything related to nuclear physics redacted and censored by military observers. "One did not dare ask a question," she recalled. At a cocktail party, she even met General Leslie Groves, he who would oversee America's secret Manhattan Project. "Groves told me that when he saw the first pile in Chicago, he understood everything in half an hour," she confided to her personal notes.[32] The "pile" refers to the first nuclear chain reaction initiated under the University of Chicago's stadium bleachers. Such a chain reaction, in which one uranium nucleus splits, and its fragments thereby convince other nearby uranium nuclei to split, and so on, was the key to either sustained nuclear power generation, or a weapon.

For Planck's part, he was long retired and, one would assume, out of the loop when Germany began its own uranium weapon efforts under the direction of physicist Werner Heisenberg. But evidence suggests his keen awareness and worry over the implications of nuclear work. The German economist Erwin Respondek was a close family friend of the Plancks, and he became an Allied spy during Hitler's regime. Planck described for him the nuclear research plans and capabilities of the Kaiser Wilhelm Society, Germany's foremost research engine.[33] In December of 1942, Planck and Heisenberg spent four days together in Budapest; it's hard to imagine they avoided the topic of nuclear research. And then, just a few months before Planck met with Meitner, he met with Pope Pius XII at the Vatican. Shortly after this conversation, the pontiff gave a public talk and cited Planck's authority: ". . . the thought of the construction of a uranium machine cannot be regarded as merely utopian. It is important, above all . . . to prevent this reaction from taking place as an explosion. . . . Otherwise, a dangerous catastrophe might occur."[34]

Presumably Planck and Meitner avoided ending their 1943 daylong conversation in the gloom of nuclear worries. No matter the final topic, theirs must have been a sad parting. She had to wonder if she would see him alive again. For Planck, Lise and her voice reminded him of much better times. Now he would make his way back to Germany, where he could await the "terrible things" to come.

4

October 1943

In the late afternoon of October 22, Max and Marga Planck pulled into the city of Kassel. Typical of his retirement years, Planck travelled about Germany speaking to audiences of scientists and nonscientists alike. Particularly during the war, he sought to raise people's spirits and keep alive the hope of science in a postwar, and presumably post-Nazi Germany.

On this particular Friday, the talk was arranged at the last minute. The Plancks were supposed to end this week in Frankfurt, but they'd been diverted because of a recent bombing there with fears of more to come. In fact, as twilight ebbed and Planck prepared a stack of notes for one of his familiar lectures, the Royal Air Force prepared another night mission. And as he began his talk that night, nearly 600 bombers grumbled aloft, heading for central Germany.

A diversionary group set a course for Frankfurt to confuse enemy defenses. But a force of 569 aircraft droned toward the smaller town of Kassel, primed to drop nearly 2,000 tons of explosives, including nearly a half-million incendiary sticks. On paper, Kassel contained military-industrial targets, including airplane and tank assembly plants. But in any night bombing, targets were approximate at best, and final Royal Air Force records show that the city center was their primary mission target.[1] Civilians were now part of the program. Echoing Planck's words to Lise Meitner, Winston Churchill spoke of turnabout as justification enough for his nation's new bombing campaign: "Those who have loosed these horrors upon mankind will now in their homes and persons feel the shattering strokes of just retribution."[2]

Despite the obvious danger of traveling during the great bombardments, Planck probably felt a deep obligation. For decades, he had provided a public voice for German science, in addresses and radio interviews. Although it is easy, even comfortable, to extend this picture of a goodwill ambassador into the war years, the historian John Heilbron notes that by *also* lecturing in occupied territory—"giving lay sermons"—Planck was guilty of ongoing collaboration with his wartime government, even long after his retirement.[3] He also consented to a Nazi propaganda film, part of Hitler's "Archive of Celebrities." The surviving films (readily available in the digital age), show him reading from prepared remarks and turning to the camera, pupils bright under a drapery of eyelids. When he faces the camera, you understand why so many students were intimidated. The hard facial angles persevered into his eighties, and his ears suggest a gargoyle's profile.

But he had a rare talent for conveying science. In his 1920 Nobel Prize address, he painted decades of physics with one gorgeous stroke: "Nothing can better illustrate the positive and hectic pace of progress which the art of experiment has made over the past twenty years, than the fact that, not just one but a great number of methods have been discovered for measuring the mass of a molecule with the same accuracy as that attained for a planet."[4]

No longer at the forefront of research, Planck believed he should contribute by lecturing outside the halls of science. A lifelong teacher, he also must have missed the classroom and its rapt, quiet faces. Lise Meitner had found his style a bit somber compared with a dynamo like Boltzmann, but most students held Planck the lecturer in awe. The young Indian physicist Debendra Mohan Bose said that Planck's lectures showed him, "what a system of Physics meant in which the whole subject was developed . . . with the minimum of assumptions."[5] Planck indeed favored this unified picture, saying that the manmade divisions of chemistry from physics and so forth, "are not based on the nature of things."[6] He remained committed to education from his earliest career moments, teaching without a position or salary in Munich, until well past his retirement, when he continued his Berlin lectures: Monday through Thursday, with student problem-solving sessions every Friday. And despite his own talents for theoretical and mathematical physics, he had pushed for increased laboratory work for Germany's secondary school students.[7]

Perhaps the best picture we have of Planck the lecturer comes from the stories of Dr. Gabriele Rabel, collected by the historian Thomas Kuhn in 1963.

> Planck's lectures thrilled me to the utmost. Yet he spoke quite simply and artlessly, with no frills, rhetoric ornaments, or jokes. He rarely cast a glance towards his listeners. When he was not busy filling the blackboard with the cabbalistic signs of theoretical physics, he seemed to read his words from

> the wood of the desk in front of him. His manner of speaking was rather strange . . . abrupt, chipped, chopped. Nor was he . . . what you would call a handsome man. . . .
>
> The fascination and thrill I felt in Planck's lectures was inspired exclusively by the unfailing logic with which he developed his thoughts, allowing us to follow every step and to share with him the joy of some surprising discovery.
>
> Only on rare occasions was this grave, methodical stream of thought interrupted by some charming, personal interlude.

By comparison, Rabel said Albert Einstein was much more personal and emotive. But in her student years (before Einstein became an international celebrity), Planck's lectures were better attended, since most students could not follow Einstein in the least. Rabel said that Planck implored his students "do ask questions," since all outcomes of a question were good. "Either you instruct yourself," he said, "or you help to improve the theory, depending on whether you are mistaken or you are not mistaken."

Regarding Planck's dedication, Rabel relayed an anecdote from a dire winter during World War I.

> We were without coal. Indeed some universities had completely closed their doors. But in Berlin, teachers and students kept on their overcoats and plodded ahead. Each day Planck made a little speech, encouraging us to endure. Once he asked us whether we wished him to give his lecture. We answered in the language peculiar to German universities, using our feet. Shuffling or scraping meant disapproval; trampling or stamping was a sign of applause. We trampled and Planck was radiant.
>
> "I rejoice," he said, "because I see from your reaction that you like to listen." The stamping increased. He added with a completely unjustified but touching act of faith: "I promise you, if you will endure today and tomorrow, on Monday it will be better."

This was the winter of early 1917 that killed many Germans far from the front lines and that eventually gave Planck himself pneumonia. The following day, a Friday, ushered in even more bitter cold. He offered again to cancel the class but was greeted by louder stomping than ever. "Full of bliss, Planck groped for words to express his emotion," Rabel recalled. "His happiness and pride at our devotion, and the childlike candour with which he expressed his joy were overwhelming, and whatever hearts had not yet flown to him before, were conquered now."[8]

In 1943, Planck was still healthy for his age, but at 85, he needed more assistance. Marga, 24 years younger than Max, accompanied her husband on his tours.

For his wartime speaking circuit, he pulled from a number of prepared lectures.[9] He often told the story of quantum theory's awkward infancy and childhood (his Nobel lecture). Far from disguising the many difficulties and his own mistakes along the way, he owned and broadcast them as a lesson in the process and agony of science. He also delivered lectures on his philosophy of science and particularly enjoyed thinking and speaking about free will and "causality," or the notion that one event necessarily leads to another, in an endless chain reaction through all time. "Religion and Natural Science" was arguably his most popular lecture, a relentlessly positive call from a religious scientist. "Religion and natural science do not exclude each other," he said. "As many contemporaries of ours would believe or fear; they mutually supplement and condition each other." They find "agreement, first of all, on the point that there exists a rational world order independent from man, and secondly, on the view that the character of the world order can never be directly known but can only be indirectly recognized or suspected." Planck argued a complementary role based on separate aims. "Natural science wants men to learn," he said. "Religion wants him to act." He punctuated this lecture with a metaphoric call to arms, as if he could render the ongoing violence of men trivial by comparison. "Religion and natural science are fighting a joint battle in an incessant, never relaxing crusade against skepticism and against dogmatism, against disbelief and against superstition, and the rallying cry in this crusade has always been, and always will be: 'On to God!' "[10]

In contrast to public response, the Nazi party disliked this lecture especially. Planck treated religion openly, without a specifically Christian flavor—he made no mention of Christ or the Bible. Indeed, Planck would later write that he did not envision "a personal God, let alone a Christian God."[11] Though he happily served as a Lutheran church elder for all of his later years, entirely comfortable with organized religion, and though he considered himself a man of deep faith, Planck appeared to hold a surprisingly similar view to his one-time friend Albert Einstein. In answering a question about his religious beliefs, Einstein once said:

> I'm not an atheist. . . . We are in the position of a little child entering a huge library. . . . The child knows someone must have written those books. It does not know how. The child dimly suspects a mysterious order in the arrangement of the books but doesn't know what it is. That, it seems to me, is the attitude of even the most intelligent human being toward God.[12]

But in 1943, as Gestapo propaganda preached to occupied countries at the time, "Neutrality towards Christ is a dangerous, even an impossible thing." Planck's inclusive religious talk may have stoked the Reich-fabricated rumors that he was in fact one-sixteenth Jewish. Planck had been worried enough about these accusations that he had his biographer detail his mother's East Prussian lineage in a second edition.[13] But the party's irritation with his lecturing grew. By the end of 1943, the Reich recommended that Planck's public lecturing come to an end.

The Royal Air Force nearly ended his speaking for the Nazis. Facing a bleak outlook in 1941, with few options to even dent the Nazi juggernaut, Britain determined to invest heavily in the machinery of bombing Germany: new airfields, munitions factories, and bomber assembly plants. By some estimates, the bombing of Germany grew to consume one-third of Great Britain's total war spending.[14] In its official 1942 embrace of the strategy, the British government cited the need to, "destroy the morale of the enemy civilian population."[15] And by bombing at night, they hoped to maximize this terror while protecting their bombers in darkness. The American air force, on the other hand, agreed to primarily daytime raids, where target accuracy was more straightforward, but their aircraft were at greater risk.

Shortly after Planck's talk concluded in Kassel, the hosts escorted Max and Marga to a bomb shelter with calm but determined experience. As they huddled together in a darkened shelter, listening for what might come next, many had to wonder again at the rumors from Hamburg. Months earlier, in late July, the Allies had incinerated the coastal town during a week of bombing. Americans bombed by day, and the British by night, cresting on the night of July 27. Incendiary bombs lit the dry city like so many fall leaves, and an unprecedented firestorm rose thousands of feet in the air, sucking oxygen into its furnace with hurricane force winds and melting the asphalt beneath. Rumors placed a death toll over 100,000, a figure beyond comprehension. Over one million refugees streamed all over Germany and the occupied lands, carrying little but for their vacant stares and sudden bouts of weeping. Though the refugees may have been loath to discuss their experience, Germans understood a new phase of the war had arrived.

Reducing Hamburg to ashes was but one part of a British focus on area bombing. From 1942 to 1945, the Royal Air Force released a million tons worth of bombs on 131 German cities, in no way limiting themselves to military and industrial targets. In fact, some suggest that if the British had specifically hit industry and infrastructure with their onslaught, they could have crippled the German state and brought the war to an earlier conclusion. Instead, over half a million German civilians died in the bombing, with over seven million left homeless by war's end.[16] Yet German industrial capacity suffered little from the campaign. In all, allied bombing reduced Germany's industrial output by just 14%, with even less of an effect on the Reich's war capacity.[17] The lack of raw materials remained Germany's primary wartime bottleneck.

Meanwhile, by all accounts, German civilian morale never suffered the intended effects. They continually cleared rubble and went about their business. Hitler ended his public appearances in 1943 but appointed Joseph Goebbels to maintain the German psyche. The minister of propaganda took on a new title, Plenipotentiary for Total War Deployment. He continually toured the bombed cities, and in 1944, he instituted a mandatory 60-hour workweek for the clearing of debris and construction of new, stronger bomb shelters.[18] If anything, many Germans seemed to hold something similar to Planck's deep embarrassment at their own fascist nightmare, with the bombing just one of the "terrible things" that must surely come their way as a measure of justice. A bombing memorial in Hamburg today reads: "The dead command us: never again fascism, never again war."

The bombing of Kassel lasted for hours, leveling the city's center. Thousands died from the blasts and subsequent fires, and about half the city's quarter million population were left without shelter. One imagines the tens of thousands huddled in their shelters as they quickly understood the magnitude of the raid. Strings of bombs would approach like so many feet of a gargantuan centipede. Planck later wrote to his son Erwin about the walls shaking. The pressure waves from the blasts left building walls flapping like sheets on a clothesline. He wrote of the many different types of bombs he learned to recognize, some unimaginably loud.

When Planck crouched in the Kassel bomb shelter, diagnosing the bomb varieties pounding above, did he think of his trench-bound sons from the previous war? Planck the father had been overjoyed when his eldest son Karl, unfocused and challenged by depression, had joined the German army in 1914. Karl trained for artillery duty, and once on the front lines, he no doubt could distinguish the 18 cm and 28 cm artillery shells flying over the trenches, and as some German soldiers reported, when a 38 cm shell exploded, "it was like meeting a monster from the sagas."[19]

Max wrote to Erwin that the air was too thick with dust and smoke; they wet handkerchiefs and covered their faces just to breathe. He took time in his letter to note that the shelter leader was especially impressive and calm throughout—ever conscious of his audience, Planck presumably realized his alarming letter needed a soothing handhold.[20] But these were the gravest moments. Once the bombing stopped and the group got to their feet, coughing and groping in the eerie whine of crumpled eardrums, they found their only exit completely blocked by debris. Max, Marga, and their newest friends were entombed under burning rubble.

The buried group somehow carved a new exit and escaped the fate of so many others that night. They emerged into what had been the streets of Kassel but were now a surrealist nightmare, lit in every hue of red and orange. Max wrote of a "sight out of Hell."[21] Survivors reported liquid incendiaries everywhere, lighting their shoes and hair. A boy carried the charred remains of his parents in a basin. A body hung from a window, decapitated by the pressurized gust from a single bomb. About 8,500 lost their lives. Most of the dead expired in basements robbed of oxygen.[22] Planck wrote that fires burned freely, with no water pressure to put them out. Parts of Kassel burned for a week (Figure 4.1).

Figure 4.1. This photograph of Kassel was taken in 1947 or 1948 by Walter Thieme, courtesy Stadtmuseum Kassel.

Max and Marga rested at a local professor's house (one of the 40% of structures to survive the night), and Planck called it the best three hours of sleep he ever had. From there, they went to stay with a niece's family, the Seidels, in nearby Göttingen.

Planck wrote to Erwin that he'd lost his pocket watch in their escape, but it was the loss of his long-time diary that troubled him most. Planck noted this diary had covered the last 10 years and so, "the experiences of these times were not really lost, but they are pushed into the darker distance." He wrote, "I'm a little too old" to start a new life, and then he directed himself to Erwin with unusual feeling. "I hope your youth provides me some invigorating fire that can renew and strengthen my aging spirits."[23]

The German writer W. G. Sebald's *On the Natural History of Destruction* attempted at the end of the twentieth century to account for the strange German silence emanating from the horrors of the bombings. For such an outsized civilian catastrophe, he argued that German literature and culture had shown an astonishing lack of digestion or even recognition of these events. "The quasi-natural reflex, engendered by feelings of shame and a wish to defy the victors," he writes, "was to keep quiet and look the other way." He notes a German woman who was observed cleaning the windows of one of the few remaining buildings in Hamburg, surrounded by devastation.[24] Sebald documents his countrymen writing much more about their fondness for the old comforts than they wrote about the actual annihilation of those comforts. Few would write directly of vignettes from Germans in a bomb shelter—for instance, "a woman with her hands closed convulsively on the Bible in her lap, an old man clutching a bedside lamp that for some unfathomable reason he had brought down with him." Instead, they often wrote in a chatty tone, "so strikingly disproportionate to the reality of the time." In the end, Sebald expresses dismay that his fellow German scribes could simply shutter themselves within, "the sense of togetherness enjoyed over *Kaffee und Kuchen*."[25] But in this way, the postwar literature of Germany simply reflected the attitudes of people like Max and Marga.

Planck ended his post-bombing letter to Erwin as he ended so many, with simple positivity. He had walked past many corpses, but he lived to appreciate the small comforts once more. "The best thing about this is finally being able to wash myself and brush my teeth," he wrote. "And then, we had a wonderful lunch, finished by coffee and cake."[26]

5

January 1944

In the depths of winter, dear friends visited the Plancks in Rogätz. The physicist Max von Laue and his wife drove from Berlin. "You can't believe (Planck) is 85 years old," von Laue had written to Lise Meitner, just months earlier. "I was astonished and admiring the clarity with which he follows the events of the times. I was very happy to see him. . . . At this age, every year is a gift."[1]

The von Laues brought with them a few of the most valuable tomes from the Planck library on Wangenheimstrasse.[2] The ongoing bombing and the damaged roof threatened the entire collection, and Erwin said they couldn't keep interlopers from roving the property any longer, as Berliners were increasingly desperate to find anything that could help them survive. The concern for the Planck home was prescient. Though earnest debates among the Allies raged about the morality and efficacy of bombing, to say nothing of the horrific casualty rate among bomber crews, the undertaking had an enormous momentum. As one U.S. military officer later said, these bombs had cost money to build and they weren't about to go to waste.[3]

Planck had been fond of von Laue from their earliest meetings as Max Laue, the student, worked toward his 1903 PhD dissertation under Planck's guidance. And the mentor may have seen himself in his student, 20 years younger. They both lived and breathed physics—not as a career but as a compulsion, a search for truth both universal and absolute. They even resembled one another, with baldness sweeping their heads identically from the front, setting off thin but expressive eyebrows and ever-present thick mustaches. But where Planck's face spoke of angles, von Laue's was more rounded, framing his large eyes, alternately intense and sympathetic. The two Maxes, one lanky with the fingers of a pianist and a preference for bowties, and the other slightly thicker with a preference for neckties, were close friends and confidantes until Planck's death (Figure 5.1).

Figure 5.1. A seated Max Planck enjoys browsing with Max von Laue in 1947. Courtesy Archiv der Max-Planck-Gesellschaft, Berlin-Dahlem.

Although he never achieved the immortal name of his thesis advisor, von Laue's career marked incredible breadth and output. In his youth, von Laue birthed X-ray crystallography, a technique that fires X-rays through crystals to determine the otherwise hidden structure within. The incredibly tiny wavelengths of X-rays provide the perfect detective tools for measuring the atom-to-atom distances within a crystal. Imagine standing outside a building without windows or doors—how could we learn of its interior? In essence, von Laue's technique gave humankind their first chance to measure the floor-to-ceiling heights and average room size from the outside, but in this case for crystals, on the scale of nanometers. This discovery earned von Laue the 1914 Nobel Prize in Physics (long before Planck or Einstein won theirs), and decades later, the technique revealed the double helix structure of DNA. He also made critical early contributions to the new field of superconductivity and Einstein's theory of special relativity. After these early landmarks, he continued a lifelong stream of important and frequently translated books.

But what few books would von Laue have brought to Rogätz from the precious Planck library? By considering the top candidates, we stand to learn a great deal about Planck's background, with its deep roots in the nineteenth century. As von Laue carries a little box of books from the back seat of his car into the Rogätz guesthouse, one catches a glimpse of what made Planck tick.

I suggest a defensible list of seven that von Laue could have presented to his friend and former teacher.

- *The Mechanical Theory of Heat* (2nd ed.), by Rudolf Clausius. 1879.
- *Popular Lectures on Scientific Subjects,* by Hermann von Helmholtz. 1873.
- *Electric Waves: Being Researches on the Motion of Electric Action with Finite Velocity through Space*, by Heinrich Hertz. 1892.
- *Critique of Pure Reason*, by Immanuel Kant. 1781.
- *The State as a Living Form*, by Johan Rudolf Kjellén. 1916.
- *Faust* (parts I and II), by Johann Wolfgang Goethe. 1808 and 1831.
- *The Theory of Heat Radiation*, by Max Planck. 1906.

The earliest physics book to have captivated Planck was *The Mechanical Theory of Heat*, by Rudolf Clausius. This tome spawned the field of thermodynamics, as Clausius searched for a more methodical way to describe the physics underlying early steam engines. *The Mechanical Theory of Heat* introduced the world to the term "entropy," and when Planck first read the second edition, it branded his supple young mind with a permanent new faith. Entropy became a lifelong lodestone for his thinking. Clausius had coined the term based on the Greek word *trope*, meaning turn or direction. The concept has bedeviled students of science ever since, proving awkward on the page and slippery in the brain. This could be due to its origin as a mathematical construct, intended to track wasted energy. Unlike some of his predecessors, Clausius rightly believed that any device using heat would always waste some of it, gone forever. He listed a precise determination for entropy: the amount of heat energy flowing into or out of an object but divided by that object's temperature. In this way, warm objects generally express lower entropy changes than cold objects, even if they absorb or transmit the same amount of heat energy. Simply put, the total *energy* in a given physics problem always sums to the same total value, and that reliable notion is the so-called "first" law of thermodynamics. Meanwhile, the total *entropy* can and does change, and that's the "second" law from Clausius.

Take, for instance, a cold mug filled suddenly with hot coffee. For simplicity's sake, let's say the mug absorbs every bit of heat lost by the coffee. The changes of energy are then equal, but not so the entropy. The cooler mug gains more entropy than the warmer coffee loses, and our pouring of coffee has increased entropy in the universe. In this way, normal everyday processes perfectly maintain the total energy involved, while often increasing the total entropy.

For Planck, it was love at first sight, and he devoted his dissertation to further exploration of the second law of the Mechanical Theory of Heat. Clausius had said that heat could never spontaneously flow from a cold object to a warm object. Rather, left to nature, heat only flowed from a warm one to a cold one. Planck further refined this sensible statement to declare something more precise and impactful: The entropy of a system would *never decrease* with the passage of time. It would increase, or in very rare cases, stay the same for a brief time. And because entropy would forever ratchet upward, Planck wrote in his thesis, "a return of the world to a previously occupied state is impossible."[4] Or, the notion that you can never go home again applies to objects and machines—not just people. This was an absolute for Planck in his early career. He imagined a world that moved ever forward irreversibly, each event in each moment the exact product of the moments before it.

When he completed his thesis, the University of Munich awarded him *summa cum laude*, but he knew better than to bask in that phrase. "None of my professors at the University had any understanding for [my thesis], as I learned for a fact in my conversations with them," he later wrote in his scientific autobiography.[5] But Max primarily awaited the comments of the father of entropy, the most respected Professor Clausius. Eventually, as young Max tired of not receiving a written reply and its anticipated praise for his work, he traveled uninvited to Bonn, knocking on Clausius's door. We picture young Planck on the doorstep, alert and smartly dressed with his hardbound dissertation under his arm, not yet 21 (Figure 5.2). But, "I did not find him at home."[6]

Planck also looked up to the legendary Berlin physicists Herman von Helmholtz and Gustav Kirchhoff. Helmholtz in particular was known throughout Europe as a brilliant polymath, moving easily between medicine, biology, and physics. When Lise Meitner at first considered becoming a doctor who would study physics on the side, her father told her that it "might be possible for a genius like Hermann Helmholtz but not for

Figure 5.2. Max Planck near the time of his dissertation work.
Courtesy Archiv der Max-Planck-Gesellschaft, Berlin-Dahlem.

another person."[7] His lifetime of work encompassed studies of hearing, sight, moving fluids, electromagnetism, and mechanics. In Planck's favorite field of thermodynamics, Helmholtz was particularly linked to the idea that energy is neither created nor destroyed but is always conserved in the universe. In time, Planck would build a friendship with this man, but not in his student days.

What did the titans of the University of Berlin think of Planck's thesis? "Helmholtz probably did not even read my paper at all," he wrote later. "Kirchhoff expressly disapproved of its contents."[8]

Undeterred, Planck immediately pursued the next academic rung, completing his habilitation, *Equilibrium States of Isotropic Bodies at Different Temperatures*, the following summer. (A habilitation was awarded after a more independent research program. It was not unlike today's post-doctoral appointment.) Of this and his other early offering in thermodynamics, he was satisfied with his progress. "Unfortunately, however, I was to learn only subsequently, the very same theorems had been obtained before me, in fact partly in an even more universal form, by the great American physicist Josiah Willard Gibbs."[9] Compared with so many influential scientists, Planck was off to a slow start. After completing his habilitation in 1880, he spent years in Munich living with his parents, teaching in a meagerly compensated low-level position, and praying for a break into the academic establishment. In keeping with a perfectionist, he shared with his college friends that, in the first year of lecturing, "I was . . . almost always unhappy with myself."[10]

Others of Planck's generation (i.e., the pre-Einstein generation), enjoyed more acclaimed and promising beginnings. In the short stack of books, Max von Laue might have also brought 1892's *Electric Waves: Being Researches on the Motion of Electric Action with Finite Velocity through Space*, by Heinrich Hertz, the most promising physicist of Planck's contemporaries. One year Max's senior, he'd generated great excitement as a student in Munich and Berlin. While Max received scientific shrugs, von Helmholtz welcomed Hertz into his research laboratories, checking on the youngster daily.[11] Planck and Hertz never overlapped. (They even seemed to orbit one another, trading student places at one point between Munich and Berlin.) But Max had an incredible respect for his work. In *Electric Waves*, Hertz described for the first time the idea of communicating electric information across a void. He reported experiments where he created intense high-voltage sparks in one device that induced smaller sparks in a completely separate device nearby. The great excitement for physicists here came from the confirmation of Maxwell's master theory of electromagnetism, which had boldly predicted such effects of electromagnetic waves years before any experiment could find them. Hertz allegedly claimed his discovery was "of no use whatsoever," and when a student asked him what would become of radio waves, he reportedly said, "Nothing, I guess."[12] Yet by 1897, Guglielmo Marconi would found The Wireless Telegraph and Signal Company, letting loose radio waves for communication. Hertz died of a blood infection on the first day of 1894, not quite 35. It was the year Max Planck pivoted to face the problem that would define him, and he leaned on Hertz's writings.

Outside of physics, von Laue might have brought *Critique of Pure Reason* by Immanuel Kant, long a Planck favorite. From Kant, Max took his fundamental views of perception, measurement, and logic. Planck, like Kant, liked to divide the world into a real and substantial realm distinct from the one processed by human sense and perceptions. Starting in his middle age, Max had written increasingly as a would-be philosopher. In particular he took up written combat with those of the "positivist" persuasion. The positivists, following their oracle, Professor Ernst Mach, believed fundamental truth to be a human fantasy; they chided scientists who aimed to understand anything that could not be seen, heard, or touched (e.g., molecules, radio waves). The idea of a "universal principle" made them roll their eyes, while for Planck a belief in the absolute was the fundamental fuel that kept science running. At the height of their publishing battle, Planck had appealed to achieving a physics that would hold for all civilizations, even alien ones. Mach snorted his reply. "Concern for a physics valid for all times and all people including Martians seems to me very premature, even almost comic, while many everyday physical questions press upon us."[13] None could raise Planck's hackles like the positivists. He wrote to von Laue promising that Mach, "will wish that he had never written it."[14] And his subsequent 1910 response to Mach surprised his friends with its resentment and personal barbs. He cast modern physicists as noble, nearly mythic creatures who must, "search above all for the enduring, the indestructible, for what is independent of human senses." He added, "That is as it always was and always will be, despite E. Mach." Planck wrote that his unworthy foe, "very often ends in error."[15]

Or von Laue might have brought his old friend one of his favorite books concerning global politics. Planck enjoyed reading authors who could look at the world of human interaction as a physicist might look at a box of interacting atoms. Oswald Spengler's *Decline of the West* was a favorite of Erwin Planck's during his most challenging and depressing years after World War I.[16] Perhaps Spengler's vision of inevitable decay as humanity's natural course comforted a family who'd lost so much and so fast. But Max Planck came to blast Spengler's later mishmash of spiritualism and science in the early 1920s, as Spengler and his co-author Rudolf Steiner argued against technology and for a new brand of science relying more on the soul than on mathematics and laboratory measurements.[17] So von Laue would have more likely brought along a book like Rudolf Kjellén's *The State as a Living Form*, the text that coined "geopolitics" and melded political realities with economics and demographics. The Swedish politician's book became widely read in Germany, both during and after World War I. Planck was known to recommend it, and at that war's end, he referenced the book in a letter to his friend Albert Einstein. He thought the book captured his own absolute allegiance to Germany, come what may. "I am eager to speak with

Figure 5.3. A contemporary mural featuring Max Planck with students at the University of Berlin. The Planck quote, shown in part, was something he in turn attributed to Goethe. "He who has come so far that he no longer makes mistakes also no longer works." The graffiti, no doubt from a student, reads, "Then total happiness!"

Photo by Dana Smith.

you about Kjellén's book sometime," he wrote before signing off. "But also about all sorts of other matters remote from depressing politics."[18]

For sentiment's sake, von Laue would have considered a favorite literary figure and plucked Goethe's poetry or drama from Planck's ruined library. Likely selections include *Faust*, or Eckermann's *Conversations with Goethe*, which had helped bring the author back to German audiences.

Max fondly quoted Goethe throughout his life (Figure 5.3). In his 1920 Nobel Prize lecture, in describing his own difficult road to the quantum theory, Planck said, "I am vividly reminded of Goethe's saying that men will always be making mistakes as long as they are striving after something."[19] And in his 1937 talk, "Religion and the Natural Sciences," he started gently, leaning on a quote from Goethe's *Faust*: "I want to deprive nobody of his sentiments and his church."[20] He would never be a scientist arguing against God.

Finally, von Laue must have brought Planck's own crowning work, a manuscript entitled *Lectures on the Theory of Thermal Radiation*, even if Max felt conflicted when seeing it again. This is the tome that, despite its own reluctance, led to revolutionary new visions for atoms, energy, and light. It unleashed a force that Planck himself did not want and couldn't contain.

The manuscript distilled years of wrestling with a family of curves and their seductive promise. We must sit with the curves because of their meaning to Planck's work and fame—it made him the worried fulcrum about which all of physics would lurch into sickening motion. As you can see in Figure 5.4,[21] the curves are simple and shapely, not unlike many that a student encounters in math and science. Each starts near zero but climbs fast, reaching a sharp summit. Lacking even a flat spot to rest, the curve launches itself to the other side, falling fast and then pulling up, gliding, and eventually skimming over the axis below. What does it mean, why was it important, and why did it, as a relatively obscure problem in the 1890s, captivate Planck?

The curves plot the brightness of light on the vertical axis, and the colors (or wavelengths) of light on the horizontal axis. A single curve describes the exact color emitted by an object at a specific temperature (e.g., picture a "red-hot" iron poker for one curve), and so the family shown in Figure 5.4 describes seven different temperatures, (listed in centigrade), with the "hotter" curves appearing more pronounced. A single curve does *not* say "it's red," but rather its shape gives an exact breakdown of a complex mix: precise amounts of each color, even in ranges beyond human vision, since objects can emit infrared light, microwaves, and, as from the sun's own version of a red-hot poker, ultraviolet.

The curves for different temperatures and even for different objects have the same essential shape (and the same mathematical form, which was Planck's stroke of genius), but they shift with changing temperature. Heat the poker? The curve grows, because the metal gives off more light.

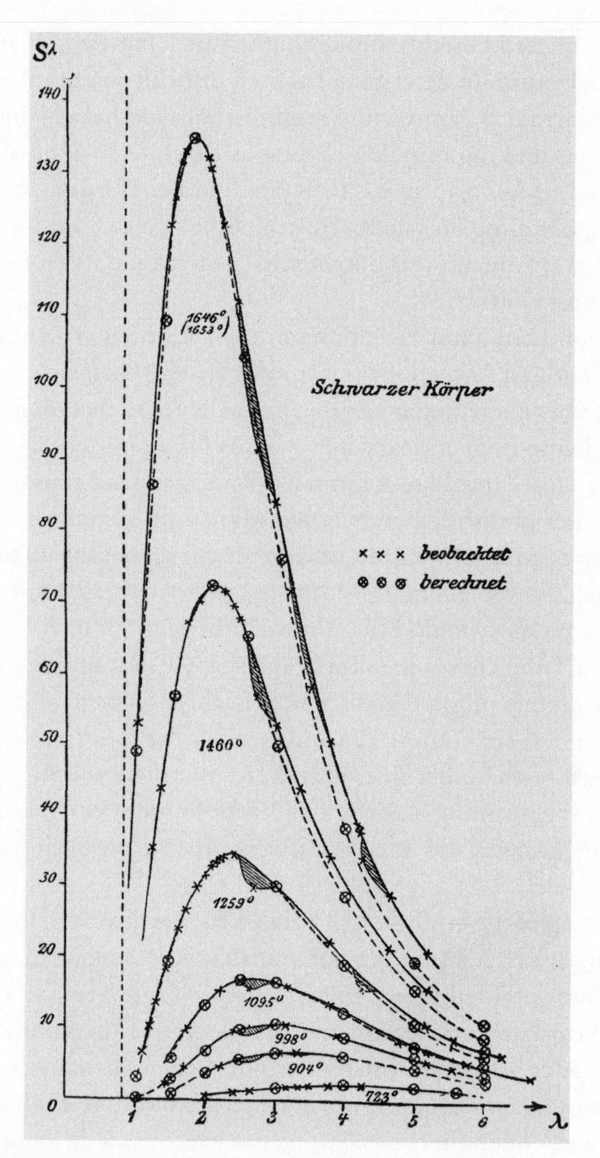

Figure 5.4. Thermal radiation data plots from 1899 from O. Lummer and E. Pringsheim. "Die Verteilung der Energie im Spektrum des schwarzen Körpers." *Verhandlungen der Deutsche Physikalischen Gesellschaft* 1 (1899): 23–41. Printed in O. Lummer, *Grundlagen, Ziele und Grenzen der Leuchttechnik (Auge und Lichterzeugung)*. Munich, 1918. (Note that this plot sets wavelength on the horizontal axis, whereas the curve in Figure P.1, for the cosmic microwave background radiation, sets frequency on the horizontal axis.)

And the curve shifts to the left a bit, meaning "blue hot" is actually much warmer than "red hot."

Taken to the heavens, these curves let us diagnose the exact temperatures of distant stars. Planck's work on these radiation curves has been and continues to be an essential tool for astrophysicists. The equation he derived to describe the curves in 1900 provides a sort of remote thermometer for scientists on Earth. Examine a star's light carefully, with more than one wavelength, and hold up Planck's equation against it, and the star's surface temperature falls into your lap without much fuss. When you look into the night sky, a certain reddish star has a surface temperature of roughly 5,000 K (8,500°F), while a blue-white star would be closer to 30,000 K (about 53,500°F).

In fact, the same curves apply to "cold" temperatures as well. The idea that an ice cube gives off light in the same manner as a star—or that an ice cube gives off any type of radiation at all—is surprising to most of us. Ice gives off much less light and it gives off light we could never see directly with the human eye, but all objects in the universe with a temperature send out thermal radiation—even, we believe, black holes. One of the young Stephen Hawking's breakthroughs was proposing so-called Hawking radiation in 1974, an exotic method by which black holes, notorious one-way "in" valves, might send out their obligatory Planck curve of light.

In the nineteenth century, the thermal radiation plotted in these curves was called "black radiation," or "black-body radiation." Note the "Schwarzer Körper," or black body, label on the historic data plotted in Figure 5.4. Completely separate from the idea of black holes, black-body radiation simply means one would observe this light curve even from a pitch-black object. So set a chunk of matter in a darkened, windowless basement. Turn off all other lights, and the emitted colors from the chunk (including those wavelengths of light beyond human perception), would be its "black-body" spectrum—the light we know emanates from the object itself, as opposed to second-hand light just bouncing off the object (which is nearly all that we see in our normal lives). In the nineteenth century, the best way to manufacture such ideal conditions was to hollow out a cavity inside a brick of material and then, through a narrow hole, measure whatever light came out. For this reason, some scientists also referred to it as "cavity" radiation. In 2015, we can contemplate with awe that the radiation curve recorded by the Planck satellite from the darkest background of the cosmos is just a cold sibling of those recorded from the little cavities of the nineteenth century—looking inward and looking outward somehow unite.

And as a few scientists peered into these cavities, Planck was beginning to flourish in Berlin. Starting with his accepting Kirchhoff's former position there in 1889, he experienced the "years of the widest expansion of my scientific outlook and way of thinking." In contrast to Kiel, Berlin surrounded him with the leading scientists of his era, and "above all the others" he formed an awestruck friendship with the great Hermann von Helmholtz.[22] So in 1894, as his mentor lay expiring from a stroke, Planck turned his attention to black-body radiation for reasons both pragmatic and passionate.

Pragmatically, the Berlin physics community of 1894 made his switch of topics obvious and even comfortable. Beyond the University of Berlin physicists, the nearby *Physikalisch-Technische Reichsanstalt* hosted a number of experimenters busily studying thermal radiation. The PTR, or Reichsanstalt in more common parlance, essentially functioned as a bureau of standards like today's National Institute of Standards and Technology in America.[23] The focus there was applied technology, so why on earth would their lab scientists look at the obscure issues of thermal radiation? They simply hoped to design a better electric light. Throughout the nineteenth century, scientists had experimented with electric lighting, first by creating bright arcs of electricity using high voltage and eventually using filaments in prototypes of light bulbs. Electric current passing through a thin filament would heat it up, and by the magic of thermal radiation (which science would not understand until the twentieth century), the heated filament would begin to glow. By the late 1800s, electric lighting became a major business opportunity. Thomas Edison began filing patents for his bulb designs, various companies in Europe and America sprang to life, and the Savoy Theater in London became the first public building with electric lighting in 1881. The key to the business was the filament material. Some filaments would burn out quickly, while others might last for a while. Some gave off sooty residue that would coat the inside of a bulb, while others glowed more cleanly. Scientists tried all sorts of materials in early bulbs: platinum, rubber, cotton, bamboo, and so on. Meanwhile, some scientists tried to better understand the exact colors and intensities of light that would emerge from a single filament at different temperatures, and this is precisely why curves like those in Figure 5.4 were measured at all.

Berlin, and particularly the Reichsanstalt, happened to house the world's foremost experts in measuring thermal radiation. These were men such as Lummer, Pringsheim, Rubens, Paschen, and Kurlbaum. Although Planck was much more of a pencil and equation physicist than these colleagues, he shared a great deal with them. They were roughly the same age, born within a decade of one another, and nearly all of them were directly tied to the greatness of von Helmholtz. Even though the lion of German science had not studied black-body radiation himself, most of these experimenters had studied under him directly for their doctoral work.[24] This collection of talented laboratory experts slowly assembled curves such as those shown in Figure 5.4 one point at a time, using evermore-clever experiments. They started in the visible realm of light, but eventually extended their experiments to look into the ultraviolet and infrared regions on either side of visible colors. Point by point, the black-body radiation mystery came into focus in the 1890s, primarily in laboratories in and around Berlin.

Planck came face to face with these experiments in his normal working circles, and as a result of circumstance. "The year '94 is a quite sorry one for my science," he wrote to friends at the time. "On January 1st, Hertz died. . . . On July 21st, Helmholtz had his stroke, which still keeps him completely incapacitated. . . . I have been quite taken up with such events, not just in my thoughts, but professionally and officially as well, and have been sometimes quite depressed by it." Before Helmholtz suffered his stroke, he had asked Planck to prepare a speech honoring Heinrich Hertz. In getting his thoughts around the amazing work of Hertz, he also considered light waves in greater detail.[25]

Planck was ready to study something different, taking his notions of entropy into a new area. He had just immersed himself in Hertz's notion of electrical waves and oscillations. And he found himself at the wellspring of all thermal radiation data, with the world's best laboratories providing constant updates.

Meanwhile, there was a lack of understanding for the origin of these curves. They needed the help of a theorist. Gustav Kirchhoff, who first described and named "black body" radiation, was dead now for many years. And every time a physicist came up with an equation for the black-body curves, new laboratory data would show that the equation was inadequate for describing the new points. For years, proposed equations fell, one after another, as the experiments continued to extend the known points on the curves.

So the need here for Planck was tangible, but nature itself now teased him. The curves whispered to him, hinting at something universal. A given curve is exactly the same for all objects and substances. To some extent, if we bring them to the same temperature and just look at their emitted light, we could not tell a hot ham sandwich from a hot ember. Imagine if every animal in a forest made the same sound each day. Chipmunks, tigers, and sparrows alike made the same cry.[26] On colder days, they made a deeper tone, and in summer, they sang at a higher pitch, but they all sounded exactly the same. Clearly, something absolutely fundamental would be at work, with some mechanism common to all species. In 1894, at age 36, Max Planck turned his full attention to figuring out *why* all objects sing the same optical song if they have the same temperature. Why are their compositions, their sizes, and their physical shapes irrelevant? The answer, he thought, would say something universal, and he suspected that his trusted friend entropy would guide him through the problem. In taking up entropy to do battle with the mysteries of emitted light, he hoped to unite the formerly distinct worlds of mechanics and electromagnetism. Black-body radiation "represents something absolute," he wrote much later. "And since I had always regarded the search for the absolute as the loftiest goal of all scientific activity, I eagerly set to work."[27] From his toils, Planck would unearth a rare gem for humanity, what his friend Albert Einstein admired as a "previously unimagined thought, the atomistic structure of energy."[28]

After thanking von Laue for the box of books and the many memories invoked, Planck presumably would have asked his younger friend to stay for a while longer. But von Laue would want to return to Berlin long before nightfall, given the danger of further bombings. If Planck walked to von Laue's car and offered one last bit of advice, out of their wives' earshot, it would have echoed what he had recently said in a letter. There's nothing to gain in provoking the Nazis. So please don't even write the name of our old friend Einstein.[29]

6

Winter 1943–1944

The Max Planck road show continued into the colder months. And despite his advice to friend von Laue, he appears to have been increasingly brave with his own words. The Swedish journalist Gunnar Pihl saw one instance (either in late 1943 or early 1944), as Planck spoke on philosophy to a group that included Nazi officers. Pihl recalled, "the little man in black," speaking, "quietly, humbly, but wisely." But Pihl was shocked by what this "gentle voice" braved to say. "He mentioned the Jew Einstein as a leader and way-shower in the world of thought, he looked beyond race prejudices and frontiers, entirely regardless of the fact that he was addressing an audience in the Nazi Foreign Office club."[1]

To mention the name of Albert Einstein in a public talk was indeed a remarkable and even risky act, but not only because the former king of German science was a Jew. He had been among the first and definitely the most famous to warn the international press about the new Nazi regime in 1933. The long-simmering resentment within Germany's far right boiled over at last, and speaking or praising Einstein became untenable. He was a traitor, a Jew who had misled German science for decades only to desert his homeland and insult it in front of the entire world. The "Deutsche Physik" movement, with its aim of erasing the damage of "Jewish Physics," reached its peak under the Nazis, but it had grown since the end of World War I under the bile-drenched care of men like Johannes Stark and Philipp Lenard. Their crosshairs were always focused chiefly on Einstein, a Jew who had found his international fame in 1919, just as the German Empire fell in humiliation.

Pihl's account is the only mention of Planck publicly praising Einstein after 1933, but Max had broadcast his feelings about the Reich a number of times. In March 1935, he gave a talk entitled "Physics Struggling for a Worldview," and Lise Meitner, in attendance, said he made his first public criticism of the Nazi regime. "The scientific irrefutability and consistency of physics comprises the direct challenge of truthfulness and honesty," she recalled him saying. "Justness is inseparable from devotion to the truth ... so too does a society of people require the same privilege.... Woe to a community when its sense of the guarantee of justice begins to vacillate."[2] Planck went so far as to decry a society in which a person's position or heritage could play any role in his legal standing.[3]

In 1935, Planck could have believed his statements might still make a difference, but by late 1943, he must have known better. So why embrace Einstein's name publicly, especially as he counseled close friends to avoid such acts? Long retired, Planck could have thought he had nothing further to lose, while those like von Laue still had years of work ahead of them. But if Planck had ever asked himself, "What can the Nazis do to me now?" he would live to regret speaking the devil's name.

Apart from simply poking the Nazi hornet nest with a verbal stick, Planck had a deep respect for and kinship with Einstein. Their shared philosophy of science overcame what many noted as polar opposite personalities—it is not inaccurate to picture the odd couple, with Planck playing Felix to Einstein's Oscar.[4] Planck was an older, conservative patriot, devoted to his Fatherland, while Einstein was a younger liberal with no particular allegiance and an eventual dream of a united world government. Where Planck was organized, stern, and subdued, Einstein was often disorganized, joking, and emotive. Where Planck dreamed of joining the academic ranks from his student days, Einstein avoided and even feared a university culture draining his precious time and independence. Both men endured significant familial heartache in their lives, but even here, Einstein's would be largely self-inflicted, while Planck's arrived on waves of unkind fate. All of these separating factors paled against the passionate faith these men shared for science.

They both *knew* that the universe operated on a set of beautiful and simple principles. They believed in pursuing an ultimate truth, and they held their *faith* in this truth's existence as their indispensable guide. In his later years, Planck finally admitted that humankind would never fully grasp the ultimate truth, but the act of forever chasing it was the most worthy of human toil. In 1929, Planck said, "Perhaps we have every reason to consider the endlessness of this continuous struggle for the prize beckoning to us from on high as a special blessing for the questioning spirit of man. It insures that his two noblest impulses will endure and take fire again and again: enthusiasm and awe." After that lecture, Einstein sent his elder a letter expressing his happy applause.[5]

Shortly thereafter, Einstein was asked to write the preface for a book of Planck's essays. At first, he asked how anyone could be credentialed enough to comment on Planck, but he agreed to try. With poignant hindsight, we can now read Einstein's 1932 preface as the last public gesture punctuating four decades of friendship. Months later, with Hitler's appointment, that friendship would rupture and fall silent.

> Many kinds of men devote themselves to Science, and not all for the sake of Science herself. There are some who come into her temple because it offers them the opportunity to display their particular talents. To this class of men science is a kind of sport in the practice of which they exult, just as an athlete exults in the exercise of his muscular prowess. There is another class of men who come into the temple to make an offering of their brain pulp in the hope of securing a profitable return. These men are scientists only by the chance of some circumstance which offered itself when making a choice of career. If the attending circumstance had been different they might have become politicians or captains of business. Should an angel of God descend and drive from the Temple of Science all those who belong to the categories I have mentioned, I fear the temple would be nearly emptied. But a few worshippers would still remain—some from former times and some from ours. To these latter belongs our Planck. And that is why we love him.

In closing his preface, Einstein turned to those who credit Planck's "energy and discipline" as the engines of his success. "I believe they are wrong," Einstein wrote. "The state of mind which furnishes the driving power here resembles that of the devotee or the lover. The long-sustained effort is not inspired by any set plan or purpose. Its inspiration arises from a hunger of the soul."[6]

Their friendship had brewed slowly starting in the first years of the twentieth century, when Albert Einstein was a student in Zurich. Einstein's life enjoys rich chronicles, including Walter Isaacson's *Einstein: His Life and Universe* and, for a narrative both more technical and more personal, the physicist Abraham Pais's *Subtle Is the Lord*. One of the most striking aspects of Einstein's career is the extent to which he approached physics and university culture as an outsider.

Born in Eastern Prussia in 1879 and raised in Munich, Einstein was already an outsider by lineage, as a Jew immersed in the passionate honeymoon period of Germany's unification. He never cottoned to life in

Munich, where many non-Jewish children teased and tormented him as he walked home from school. He especially recoiled from the rampant militarism surrounding him. During his secondary schooling, he fled the regimented environment of Munich in favor of the Italian Alps. At the time, as Max Planck began his investigations of black-body radiation, a 15-year-old Albert Einstein declared to his parents that he would not be returning to Germany.[7] Instead, he would try to pursue college in Zurich directly.

Most of us enjoy the stories of Einstein's early struggles with learning, but they are largely overstated. It's true that he spoke rather late and was known as "the dopey one" in his family. He liked to practice phrases to himself before speaking them aloud and continued this pattern for most of his life, when he found phrases or words that delighted him.[8] But Einstein was regularly first in his class, and he particularly excelled in mathematics. "Before I was fifteen I had mastered differential and integral calculus," he later recalled.[9] By this age, he was also reading the philosophy of Immanuel Kant for fun, and Kant remained one of Einstein's (and Planck's) favorite thinkers. Unlike Planck, Einstein was not always diligent and disciplined as a student. His mathematics professor Hermann Minkowski labeled him a "lazy dog."[10] The professor eventually joined his former student at the forefront of Einstein's new physics but never amended his statements about Einstein's university days.[11]

After graduating undistinguished from the Zurich Polytechnic Institute in 1901, he pursued PhD studies with little success. Meanwhile, he took work at the Swiss patent office, where he stayed for eight years. With evenings consumed by physics, these years formed the most creative and intellectually diverse of his life. He continued to pursue a doctorate degree, though most of his ideas were too radical for his professors' tastes. Even here, his goal was not to find his way to a professor appointment, since he had no fondness for universities. For one seeking the true nature of the universe, accepting a post at a university would be selling out. Even when he finally took a professor appointment in 1909, he wrote to a friend, "So now I too am a member of the guild of whores."[12] But even if Einstein entered the temple of science through a side door, he did so with sincere awe. He eventually appreciated in Planck what he knew in himself: that a real scientist wanted only to know the mind of God.

Planck and Einstein had first corresponded during the first years of the twentieth century, when Planck was in his early forties and Einstein in his early twenties. Unlike the era of Planck's youth, when Professor von Jolly told him the foundation of physics was nearly complete, the edifice presented major cracks to young Einstein. As he eagerly wriggled his fingers into every crevice he could find, Planck encouraged him. Einstein published his first papers in the *Annalen der Physik*, then the most prestigious journal of physics. Planck was not the chief editor at the time, but he was typically charged with handling the more theoretical and mathematical submissions. Planck was a major voice on the journal's editorial staff for 45 years, and as the historian Dieter Hoffmann puts it, "Max Planck should be regarded as the *Annalen's* 'strong man,' if not its 'grey imminence.'"[13] We can reasonably presume that Planck was one of the first, if not *the* first, to read Einstein's scientific offerings.

As an aside, Planck's sterling work as an editor provides perspective on science then versus science now. Editor Max handled papers in a manner that puts modern print journal standards to shame and would make even our online journals blush. Without so much as a typewriter, Planck would be in touch with authors within two days of their submissions, and the author of an accepted paper could expect page proofs (the version immediately preceding publication, in which an author checks for typos) within a week.[14] To be fair, the volume of submissions has increased exponentially in the last hundred years, and Planck could serve as judge and jury in most cases, as opposed to finding three reviewers.

As of 1900, Einstein was revving up for a run of theoretical publishing that would be worshipped in any era and by any standard. His first five publications in *Annalen der Physik*, while not pedestrian, were not technically earth-shaking, and having gained nothing significant, he was reportedly not happy with the results. The next three in 1902, 1903, and 1904 were more substantial, helping bolster the infant field of statistical mechanics.

At age 26 and still without a PhD thesis, Einstein forever changed science in his "miracle year," 1905, with five works that were as incredible as they were diverse. He made stunningly accurate computations of molecular sizes, substantially expanded Planck's own black-body work to other arenas, and perfectly described the chaotic motions of tiny particles under a microscope (so-called Brownian motion). In the last of the papers, he coined what has become the greatest scientific cliché, the equation of mass (m) and energy (E) equivalence, relating the two via the square of c, the speed of light. But one of the five papers stood out in particular to Max Planck. It launched the special theory of relativity and eventually made Einstein the de facto ringleader of "Jewish Physics" to the Nazis.

This paper, the fourth, proposed a radical yet simple idea for the behavior of light in the universe. Although disarmingly straightforward, the new thinking had deep implications for the way we understand space and time. Einstein's ideas didn't emerge out of the ether (so to speak), but rather to address a significant problem in physics.

Debating the exact nature of light had a rich history by this time. According to Newton, author of 1704's *Opticks*, light was composed of particles. This is known as a "corpuscular picture," where light is made of billions of little bullet-like entities, flying from the light source and bouncing off objects and arriving in our eyes to let us see those objects. But experiments of the early 1800s showed strikingly *wave*-like properties for light, and for most of the nineteenth century, the wave theory of light dominated scientific discourse. In this view, light was an oscillating disturbance, like a rolling water wave on the ocean.

The vexing problem with the wave theory involved the question, "Waves on *what* exactly?" Other waves were known to need a substance, like water for ocean waves, or air for waves of music. On what exactly did light move? The answer was widely agreed to be the luminiferous aether (or simply the ether), an invisible but hopefully, real substance that filled what would otherwise have been disturbingly empty space. The foggy notion of ether predated even Newton.

A comforting but somewhat hidden thought lurked in the ether: an absolute type of space. Once we define a substance that fills space, it gives all future conversations a home base, or a reference point. It speaks to a tacit human craving: absolute stability, or a starting point from which to reference any errand or journey. But significant problems arose for the ether when scientists tried to find it. While everyone assumed that the Earth moved through the ether on its orbit around the sun, all experiments failed to locate the "ether wind." Imagine moving in what you were sure to be a very fast car, but when you roll down the window and extend your hand, you feel no wind whatsoever. You run a test with two ping pong balls, throwing one in the forward direction and one in the backward direction. Though you expect to notice a difference, especially for the ping pong ball thrown into the wind, you measure no difference at all. You might conclude that the air was being "dragged" along with your car or that your car wasn't really moving as fast as you thought. Or you might question the very existence of air. And this was the type of confusion for physicists at the end of the nineteenth century. They witnessed no sign whatsoever of an ether wind.

Planck himself became intensely interested in late 1898, despite being in the throes of his consuming black-body investigations—or perhaps as a break from his frustrations with the problem. He had just met a brilliant physicist from the Netherlands, a kindly, middle-aged man named Hendrik Lorentz. After decades of hermit-like work, Lorentz had attended his first conference, where he and Planck fell into conversation about the ether problem. Planck wrote to Lorentz that their meeting was, "one of the most precious results of the Düsseldorf conference."[15] They then exchanged a half dozen letters about the ether problem through early 1899. Max was among those who thought that a layer of the ether must surely adhere to the Earth and move with it through the cosmos. The two men ironed out some mathematical differences of opinion, Planck published a lone paper on the topic, and then he was satisfied. "I myself am not going to return to this issue," he wrote to Lorentz. "I leave this whole theory to your leadership."[16] In retrospect, this detour with Lorentz primed Planck's later advocacy for a new perspective.

In 1905, young Einstein offered a bold solution by focusing on what *united* all different systems, no matter how they moved with respect to each other. In Einstein's proposal, the ether was inconsequential. At first he referred to his new ideas concerning light and motion as "invariance" theory. And indeed, the foundations of what we now call special relativity are not about anything relative at all. Soon, young Einstein heard from an enthusiastic fan. "My papers are much appreciated and are giving rise to further investigations," he wrote to a friend. "Professor Planck has recently written to me about that."[17] Max Planck, as a journal editor, was among the first people to see Einstein's new paper, and where most others had doubts, Planck saw brilliance. He was certainly the first physicist to hold it aloft and spread the word, calling it "relative theory." Others, like the young physicist and Einstein friend Paul Ehrenfest, took to calling it "relativity theory." This name may well have added to confusion that understandably persists for broader audiences.

When Einstein later became a celebrity, every household in Germany wanted better access to the meaning of relativity. Despite Einstein's own marvelous book for the general public, misinformation flourished. A popular short movie of 1921 aimed to "explain" relativity in terms everyone could understand. It showed a boy in a classroom, agonizing over the last 15 minutes of class. Then it showed the same boy later, talking with a girl, saying now 15 minutes was "so little!" Voila! The film claimed: Now you understand relativity.[18] Such mutilations of his work frustrated Einstein to no end. Planck tried to set the record straight, even in his scientific autobiography.

> The often-heard phrase, "Everything is relative," is both misleading and thoughtless.... Our every starting-point must necessarily be something relative. All our measurements are relative. The material that goes into our instruments varies according to its geographic source; their construction depends on the skill of the designer and toolmaker; their manipulation is contingent on the special purposes pursued by the experimenter. Our task is to find in all these factors and data, the absolute, the universally valid, the invariant, that is hidden in them. This applied to the Theory of Relativity, too.[19]

Nevertheless, the name that stuck emphasized the crazy-making differences between observers in relative motion, instead of what Einstein made immortal and constant: the speed of light and the laws of physics.

Einstein built his theory by setting those two notions on a pedestal, and then watching, with a merciless logic, where they took him. These assumptions resonate with the idea that the universe has an underlying simplicity and beauty. (For both Planck and Einstein, the beauty of nature was independent of whether or not humans could discern it or even stomach it.)

The first assumption is an egalitarian one that would seem completely noncontroversial: The laws of physics will be the same in any "inertial reference frame." (A reference frame means, in essence, an impartial observer and her little set of physics equipment; inertial simply means that she and her small laboratory are either at rest or moving at a constant speed.) Whether we test the laws in a train at rest, or in a train that is zipping down the rails at 100 mph, we find that the same laws of physics hold firm. According to Einstein, they are so much the same that if your train is quiet and you draw the shades closed, you should not be able to tell if your train is moving or not, just by running physics experiments within.

We have to pause here and point out that this first principle has deep roots. In the seventeenth century, Galileo reasoned that a sailor cloistered below decks on calm seas couldn't know for certain if his ship was moving or not. If the portholes were closed, and his ship moved at a constant speed, in one direction, he would observe the same basic physics as someone sitting on terra firma, watching the silent ship sail past. This concept was revolutionary at the time. Galileo had done away with the idea of absolute velocity, saying instead that all velocities must be referenced against *something* to have any meaning at all (like a mountain for an Earth dweller, or like a star for a space traveler).

But Einstein's second principle took Galileo's notion into a brave future. The second statement of special relativity declares exactly what measurements suggested: "Light is always propagated in empty space with a definite velocity *c* which is independent of the state of motion of the emitting body."[20] So in Einstein's proposal, the results of the various light-measurement experiments were to be expected. No matter how the Earth turns against its background of space (or ether), the speed of light will always be the same. This too sounds fine at first glance, but in short order it starts to cause problems for what the human mind and experience call "normal."[21]

In proposing just those two simple postulates, Einstein quietly threw much of what humans find comforting under a rushing and merciless bus of physics. The ensuing complications, counterintuitive to human experience, would have left even Galileo stunned, and they absolutely stoked a backlash against special relativity's author. The blowback was too violent to stay on the scientific page, and it even came to threaten Einstein's foremost defender, Max Planck.

One of Max Planck's most incredible and rare attributes was his ability to open and change his scientific mind, even at 47, which is, by time-honored stereotype, a somewhat stodgy age for physicists. He embraced the theory immediately and never looked back, even while the majority of physicists, including titans like Hendrik Lorentz and Henri Poincaré, struggled to accept it. How could mental complications (let alone animosity) arise from two such innocent postulates?

First, one can point to the loss of any truly fixed point in the universe. We noted earlier that most of us unconsciously assume that there is some "home base." If the Earth moves with respect to the sun, that's a start, but the sun also moves with respect to other stars. There must surely be a center and an origin: a place at absolute rest. In a very real sense, the luminiferous ether of the nineteenth century was an attempt to grasp and retain this sense of stability. But special relativity happily said there could be no center, no anchor in this universe.

Two other quiet assumptions had fixed themselves in the human mind before 1905: absolute length, and absolute time. If Karl stands next to a moving train, and Emma rides on the moving train, classically we don't even need to state the obvious: Both should agree on the length of the train—it's 300 feet long—and both should agree on the amount of time elapsed as the train rumbles past Karl—10 seconds.

But Einstein saw that preserving these comforting notions would mean that the laws of physics could not be uniform, and he saw that such a world would have different measurements for the speed of light, based on relative motion. Following his postulates instead of comfort meant the end of absolute length and absolute time. Indeed, in the train example, the two observers do not measure the same length of the train—Karl sees it as shorter than passenger Emma does. Nor do they agree on the time elapsed as the train passes. And if that isn't bad enough, here is what induces nausea in physics students every year: Both observers are correct.

Particularly troubling to the human mind is the loss of simultaneity. If two observers move with respect to one another, they no longer agree as to what exactly is simultaneous. If Karl, standing by the train tracks, extends his arms, and snaps his fingers simultaneously with both hands, Emma, moving past at a very high speed, will not agree that the snaps are simultaneous. Literally, in her "reference frame," one snap precedes the other.

Although Einstein's new theory baffled most physicists, Planck celebrated it. Within months of publishing Einstein's first relativity paper, Planck brought the topic to his university's winter colloquium series, even correcting a tiny math error in Einstein's paper.[22] In 1906, Planck published the first follow-up to Einstein's new theory, showing that his own beloved entropy, as well as the principle of "least action"—a mathematical statement of nature's efficiency—was preserved in the world of special relativity. In short, while the new theory confronted physicists with confusing paradoxes and the loss of long-held assumptions, Planck worked to comfort the world, showing that some sacred physics cows grazed unmolested.

Einstein had generally been disappointed in the response to his incredible 1905 work, but he delighted in the attention from Planck. When the physicist Walther Kaufmann attacked the $E = mc^2$ proposal in 1906, saying his measurements showed that mass had no dependence on an object's total energy, Planck defended Einstein publicly, saying Kaufmann's experiments lacked the precision required for such claims.[23] And when Planck's 1906 follow-up relativity paper drew criticism, Einstein wrote expertly in defense. Planck sent an appreciative note in the summer of 1907. "As long as the proponents of the principle of relativity constitute such a modest little band... it is doubly important that they agree among themselves."[24]

In summary: Let's stick together, because we're right. The brainy pen pals had still not met in person. Einstein later said the main reason other physicists gave relativity a chance was "the decisiveness and warmth with which [Planck] championed this theory."[25]

The reason that most human beings struggle to accept special relativity is very simple. Because we do not encounter speeds fast enough to notice its effects, we're left to trust the strange claims of equations.

The actual effects of special relativity depend on algebraic terms that compare the relative speed of two observers divided by the speed of light. Because the speed of light is so enormous, such factors are normally so small that we would never notice them. They truly emerge as relevant when the relative speed of objects approaches 1% or more of the speed of light. At present, we have no trains planned that will travel at .01 *c*, or 3,000 kilometers (from Boston to Houston), in one second. And even if a commuter train moved at this speed, the effect of special relativity on two synchronized watches (one left at the station and one moving on the train), would be tiny. If the train ran for an hour at its incredible speed, the watches would appear to be off by just a quarter of one second.

The mental chafing we experience with special relativity can be soothed with a cognitive analogy, in terms of physical challenges to our deeply held intuitions. In 1842, the Austrian physicist Christian Doppler proposed the now-famous and well-accepted effect that bears his name. He proposed the observed shifting of frequencies for astronomical observations (for which the effect has been revolutionary, from discovering the universe's expansion to presently helping us ferret out hundreds of planets orbiting distant stars), but he also mentioned that it should happen for sound waves on air.[26] And in modern life, the Doppler effect is so familiar that we take it for granted. We hear it nearly every day in the changing tone, weee-EEE-OOO-ooh, of a passing car or siren. The tone depends on whether the siren is standing still, moving toward us, or moving away from us, and the speed is important. We expect to hear it from a fast-cruising motorcycle but not so much from a child's slow bicycle. The Doppler effect is nothing that makes us stop to marvel, at least as adults.

But for a moment consider Max's father, Wilhelm Planck, attending a symphony concert in 1841. Imagine a violinist playing a long C note as he prepares for the performance, just as Wilhelm walks into the concert hall. Would Wilhelm assume that he hears exactly the same note as someone already seated, or someone exiting the hall? Of course. But what about for a person running very quickly into or out of the hall? Again, Wilhelm would assume that a C is a C is a C. Any other suggestion would be nonsense to him. We now know that either a moving source of sound, or a moving receiver of sound, will shift the received pitch via the Doppler effect. If we moved into the concert hall fast enough, the long C note would be shifted to C-sharp. If we moved *away* from the concert hall fast enough, the C note would be shifted to C-flat. But this wouldn't have been conceivable to Planck's parents, despite their intelligence and careful listening. They simply hadn't had the chance to hear the effect manifest in 1841, since they were never moving fast enough.

Shortly after Doppler's 1842 presentation, his effect was confirmed by having a band play on a moving locomotive platform. Because transport was becoming ever more rapid in the mid-nineteenth century, people had opportunities to notice the pitch shift. Old tacit assumptions—every listener hears exactly the same pitch, no matter their motion—eroded under a storm of new experiences. We accept now that an ambulance driver doesn't hear his siren's pitch shifted, even though we do as he passes.

Conceptually, it is fair to compare that cognitive process to the results of special relativity. We have little reason to question the length of a ruler or our synchronized time pieces because all the objects or vehicles among us move at ridiculously slow speeds compared with that of light. Assuming an absolute like "a violin note is a violin note, for all possible audience members" is analogous to assuming "a ruler always measures 12 inches, no matter who looks at it." Doppler, in his way, saw that the *real* absolute was the frequency of the sound-making object (as opposed to the *received* frequency). Einstein recognized a different absolute when he made the speed of light immutable, while necessarily opening up aspects of space and time to flexibility. For now, we can only "hear" the effects of relativity by listening through equations, but few of us have both the flexibility and the mathematical faith of Max Planck.

After reading Einstein's 1905 paper, Planck threw himself into relativity work, even while he dismissed Einstein's new paper that expanded his own 1900 quantum breakthrough. He recognized that relativity would have direct implications for Newton's mechanics, a cornerstone of physics. By 1907, Planck had derived a new relativistic dynamics, since old quantities like momentum and energy required a significant makeover at high speeds. Along the way, he constructed a new and compelling way to justify

Einstein's $E = mc^2$. Thirty years later, Lise Meitner used this expression after her snowy walk in Sweden to compute the power unleashed by the splitting (and mass reduction) of a nucleus.

Beyond serving as a special relativity champion in the scientific community, Planck would spend the better part of three decades sticking up for Einstein in public as well. With Einstein's fame came attacks from those who decried relativity as false prophecy. The nonintuitive and even disquieting nature of relativity helped it achieve pariah status in the eyes of the far right. Even today, some fringe scientists claim that physics went off the rails in 1905. They claim that Einstein's relativity and Planck's quantum theory cannot possibly be right, since these theories speak against common sense. (If a reader harbors any doubts, special relativity has been confirmed time and time again, using dozens of methods.)

To many, Einstein and Planck were moving to take physics away from where it belonged: in the laboratory. According to physicists like Johannes Stark, a physics built by pencil and paper, instead of measurement, was pure fantasy. When such attitudes flowed into the returning tides of anti-Semitism, they brewed a nefarious conspiracy theory called "Jewish Physics." As the principal architect and main Jew behind it all, Einstein would come to fear for his life, long before the Nazis came to power.

Einstein would finally reply to German anti-Semitism by rejecting his uncomfortable homeland. With Hitler's rise in 1933, Einstein publicly severed his ties with Germany and German science in particular. "I have never had a particularly favorable opinion of the Germans (morally and politically speaking)," he wrote to a friend. "But I confess that the degree of their brutality and cowardice came as something of a surprise to me. . . . The lack of courage on the part of the intellectual classes has been catastrophic."[27] Still, as he wrote to the exiled Fritz Haber, he excepted a "fine few personalities," namely, Planck and von Laue.[28] While von Laue continued teaching special relativity to his German students in 1933, even deadpanning that the theory had been written in Hebrew, Planck stood awkwardly with Hitler's new government, even as it condemned Einstein.[29] But the odd couple somehow parted with their mutual respect intact. Planck had requested as much in his last letter, and Einstein gladly complied. "In spite of everything, I am happy that you greet me in old friendship and that even the greatest stresses have failed to cloud our mutual relations. These continue in their ancient beauty and purity, regardless of what, in a manner of speaking, is happening further below."[30] The old friends never communicated directly again, though they continued to speak well of one another.

A decade later, with the world at war, Planck spoke of Einstein at his own risk. The journalist Gunnar Pihl recalled his winter evening at Planck's talk with a sort of awe.

> It was like being present at a ceremony and a sermon. A violent contrast to the spirit of the place and the egotistical vociferousness of the Nazi ministries. Everybody felt moved, . . . everybody except the few party potentates in the room. They could not grasp what had happened to the others and did not dare to ask the reason for the atmosphere of reverence.[31]

7
February 1944

Some Allied voices now hoped to end World War II with intensive bombing of German cities. Britain's Air Officer Commanding-in-Chief Arthur "Bomber" Harris decided Berlin must particularly suffer. "It will cost us 400 to 500 aircraft," he said. "It will cost them the war."[1] Britain planned to weaken Germany's will, perhaps to the point of surrender, by dropping months of kilotons on Berlin. The aerial Battle of Berlin lasted from November of 1943 through March of 1944, and the city recovers under a busy set of construction cranes to this day.[2] In hindsight, despite catastrophic damage, the mission failed completely, but Harris and his obsession with pure destruction survived all internal criticism.[3]

On February 15, 1944, nearly 900 aircraft left Britain on a clear Tuesday night and assembled in formation over the North Sea. Unleashing arguably the most brutal air raid on Berlin, the mission aimed to terrorize the city's center and to level Siemensstadt, an industrial park west of the city.[4] In accomplishing both, the bombers also demolished a number of western suburbs. If the Planck home could have survived this night, it may have survived the war intact, but direct hit after direct hit pulverized and incinerated the homes of Planck and his former neighbors.

Max and Marga had longed to simply repair the roof at Wangenheimstrasse 21 and resume their former habits, but the news of February 16 abruptly ended that fantasy. Erwin's wife Nelly poked through the ruins of the Planck home looking for anything to salvage.[5] The only books from his library to survive were those few von Laue had carted to Rogätz weeks earlier. The one night raid erased from history decades of Planck's correspondence, notebooks, files, and journals.

Planck allowed himself no time to despair. He thanked Nelly for her attempts and said working in Rogätz suited him. Even then, he put the last flourishes on a new lecture, "Phantom Problems in Science," a new assault on scientific pessimism.[6] We have this lecture now in its form delivered after the war, when Max spoke to an audience in Göttingen. By "phantom" he means that a key part of the would-be problem or question is either poorly constructed or unrealistic from the outset. Douglas Adams provided perhaps the best example of a Planckian phantom problem in his comedic novel *Hitchhiker's Guide to the Galaxy*. A planet-sized supercomputer is asked to render the ultimate answer to "life, the universe, and everything." After calculating for many hundreds of years, the computer has a simple but enigmatic answer: 42. When the tired computer then informs the confused and enraged masses that, instead of raising their fists to the answer, they should come up with a more logically worded question, Max Planck would have applauded. He considered many conundrums of philosophy (free will vs. determinism) and quantum physics (wave vs. particle) as symptoms of inappropriate or poorly worded questions.

In the closing passage of "Phantom Problems," Planck made a striking statement about "one's own conscience." It is not clear whether he wrote this while still under the thumb of the Nazi regime or added it after the war, with more painful hindsight. "Under no circumstance can there be in this domain [one's own conscience] the slightest moral compromise, the slightest moral justification for the smallest deviation. He who violates this commandment, perhaps in the endeavor to gain some momentary worldly advantage, by deliberately and knowingly shutting his eyes to the proper evaluation of the true situation, is like a spendthrift who thoughtlessly squanders away his wealth, and who must inevitably suffer, sooner or later, the grave consequences of his foolhardiness."[7]

Most would assume Max saw himself as worthy of those high standards, but given his compromises of the Nazi years, and the subsequent loss of all material possessions, one could also ask if he flogged himself here. Was he admitting to "shutting his eyes," and then suffering from his own "foolhardiness"? Could he have written these lines without imagining the ruins of his long-time home and his happiest times?

A legion of possible memories of Wangenheimstrasse 21 could have haunted Max Planck: seeing his designs for the home realized in brick and timber; moving into the professor-filled community with his first wife Marie and their pack of teenagers; assembling his library with systematic devotion; concentrating and writing there in the early, chilly mornings; running and winning a game of tag with young Lise Meitner or Max von Laue in the yard; winking and making another serious "Tarok" card play;

enjoying a late cigar with his neighbor and friend, the theologian Adolf von Harnack. Most of all, Max would have mourned the end of his family's musical evenings on Wangenheimstrasse. The onslaught of February 15 silenced the last echoes of so many concerts. And it buried in rubble the beautiful singing of his late wife.

In his Munich days, Max had a classmate by the last name of Merck, and this classmate had a musically gifted sister named Marie Eugenie, three years Planck's junior.[8] The Merck family was affluent, and most probably more so than the Plancks. Marie's father was a successful banker, heading Merck, Fink, and Company.[9] Did Marie get to hear Max playing the organ at their Lutheran church? Did he attend a recital of hers at the Merck home? Was his college operatic composition, "Love in the Forest," written and performed with Marie in mind? We don't know.

Whatever Planck thought of Marie, he didn't propose to her in the six years he lived with his parents after completing his doctorate, waiting with increasing anxiety for his first real academic job. Despite his diligence and excellent work, finding work as a theoretical physicist (as opposed to an experimental physicist) was difficult. In April 1885, as he approached 27 and Marie was not yet 24, his luck turned and he accepted an appointment as Associate Professor of Mathematical Physics at the University of Kiel.[10] "This offer came as a message of deliverance," Planck wrote in his autobiography. He received the official offer in a Munich hotel. "The moment … he informed me of the particulars and conditions of my appointment, was, and will always be, one of the happiest of my life. For even though my life in my parents' house was as beautiful and contented as any man could wish for … I was yearning for a home of my own."[11] Planck only obtained this job offer through a stroke of luck. The faculty had originally voted to offer the post to the amazing Heinrich Hertz, but while the local government's budget process delayed an official university offer, Hertz took a job in Karlsruhe.[12]

Planck's proposal to Marie Merck may also have been delayed by a lively group of friends. He enjoyed a close-knit group in his college days and beyond. Before taking a college year in Berlin, he toured through Italy with a few of these guy pals, with visits to Venice, Florence, Genoa, and Milan. The four established a *Brieftagebuch*—literally a "letter diary"—a jointly curated journal, functioning as a long-form collective letter to one another. Each would keep the book for no more than two months, detailing their thoughts and experiences, and then they would send it to the next in line.[13] Carl Runge, a brilliant mathematics and physics student, was an avid participant in the letter diary. According to Runge's daughter Iris (who was also his biographer), Runge and Planck in particular fell into long philosophical conversations, and Runge liked to push the more conservative Planck's comfort zone on the topic of religion, "e.g. whether the Christian church overall had been more detrimental or beneficial." Apparently, Planck refused conversations of that sort.[14] And while the completely Prussian Planck was still notably stiff, even in these informal pages, he did try his hand at teasing his friends. In chiding Bernhard Karsten for a very short offering, he suggested, "a minimum number of rows or facts shared" per entry.[15] In all, the incredible project ran in fits and starts from 1878 to 1927 (the year of Carl Runge's death), and one is treated therein to literally hearing Planck and Runge age on the page.

After the trip through Italy, Runge accompanied Planck to Berlin for a year of mathematics and physics courses, including lectures by the physics titans Helmholtz and Kirchhoff. One can almost see the youngsters poking one another in class there, because, despite the eminence of both physicists, they lamented Helmholtz's disorganized mumbling and, simultaneously, Kirchhoff's overly organized lectures, which felt dry and robotic. Planck noted that Helmholtz, in particular, "repeatedly made mistakes" and "his classes became more and more deserted." Planck was one of the very few to keep attending.[16]

Planck's entries after accepting his first job, in 1885, illuminate his bachelor life in Kiel. He lamented the lack of music in the small town, saying there was no comparison to Munich, but in the summer, "almost every day I come to the water, next time hopefully *into* the water." And he spoke of a new group of friends, boldly calling themselves the "society against overpopulation" with engagements, marriage, and child rearing "strictly taboo."[17] This appears to have been, for most, a tongue-in-cheek game, and a quieter announcement followed soon enough.

Max admitted complete defeat in the fall of 1886. "We are now a couple," he wrote, with no mention of his fiancée's name (Figure 7.1). "The commonly inevitable fate has befallen me as well, and so quickly, that I can thank you today for all your best wishes, which have really helped. One cannot expect much said from a groom; he has to deal with so many new and unexpected things going on inside him that it will take a while before everything has returned to order, before he can emerge from the shadows." Emerging from that pall of romance, Planck closed his engagement

Figure 7.1. The newly engaged Max Planck and Marie Merck, 1886.
Courtesy Archiv der Max-Planck-Gesellschaft, Berlin-Dahlem.

announcement, before mailing the *Brieftagebuch* to its next host. "In the winter I will teach mechanics. With best regards to you all, your faithful Max."[18]

Runge wrote the next merciless entry just two days later. "Max is still really good . . . he wrote to us that he belonged to a club in which engagement is strictly frowned upon. . . . Hardly are the vacation days here, so he has nothing more urgent to do than act on tendencies diametrically opposed. I am waiting to hear what penalties the club will impose on him."[19]

Max and Marie were married on March 31, 1887, with Max approaching 29 and Marie 26. They wed in Munich, just a few days after his brother Adalbert, now an engineer, married Johanna Baur.[20] Max and Marie moved into a Kiel sublet, and he resumed his lecturing and writing.

In short order, one birth and two deaths would have an incredible impact on Planck's arc. Karl Wilhelm Hugo Planck, the first of four children born to Marie, debuted in Kiel on March 9, 1888. Karl was born into a simmering time. On the day of his birth, the German Empire lost its first Kaiser. Wilhelm I passed away, launching "the year of the three emperors." Heir apparent Frederick III took over, but with a raspy voice. Losing a fight with cancer of the larynx, Frederick's reign expired in June, when ambitious Wilhelm II took the throne. Though a young 29, he was anything but liberal, and some would say that Germany's course to World War I started in their new ruler's valiant visions of conflict.

Meanwhile, months after Marie had moved to Kiel, Planck learned that his former professor Gustav Kirchhoff died in Berlin, setting the university to search for a replacement. In 1888, he agreed to edit and publish a posthumous compilation of Kirchhoff's work, and he also expressed his interest in the Berlin vacancy. Aside from the lack of music and culture in Kiel, teaching physics in a small town was proving frustrating as well. "The courses this summer are horrible," he wrote to the letter diary that summer. One of his classes had been cancelled and another had only two students in it. "And the worst part is that these bad times will probably last quite some time."[21] (In truth, he may have meant for physics in general, and not just in Kiel, as the subject was still not very popular.)

Although Berlin was impressed with Planck, and von Helmholtz had taken a liking to the young physicist, Berlin had two other intriguing finalists for their job opening: Heinrich Hertz (he who had just discovered electromagnetic waves), and Vienna's Ludwig Boltzmann (he who believed in atoms, and who would later impress Lise Meitner with his lectures). The University of Berlin ranked Planck third in this trio, but they turned to him after Hertz stayed with his busy experiments in Leipzig and the mercurial Boltzmann turned them down. (Boltzmann would later tell Meitner this was among his life's greatest mistakes.) Planck was all too ready to accept, making it official in late 1888, and then moving to Berlin in the spring of 1889.[22]

Meanwhile, as Max worked out the details of his new post, Marie was already pregnant with twins. Poor Marie uprooted her household in the spring of 1889 either massively pregnant or holding a new baby girl in each elbow. She bore Emma Ottilie Agnes Planck and Margarete (Grete) Elisabeth Casimira Planck (Figure 7.2) on April 10 (just 10 days before the Hitler family of Branau, Austria, welcomed baby Adolf).

Figure 7.2. Emma, Karl, and Grete Planck as children.
Courtesy of Archiv der Max-Planck-Gesellschaft, Berlin-Dahlem.

So with a chaotic home and presumably with bags beneath his keen 31-year-old eyes, Planck prepared for his first major talk in Berlin. His most recent calculations dove into a newly uncovered body of chemistry. He took the mental tools of thermodynamics, including his favorite entropy, to the world of electrolytes, liquid solutions packed with ions like sodium and calcium. These were of great interest for their ability to conduct electricity and for their implications for the emerging picture of atoms. In this field, Planck swam with some emerging leviathans of "physical chemistry," like Sweden's Svante Arrhenius and Amsterdam's Jacobus van't Hoff. In 1889, Arrhenius had just proposed the intriguing new idea of "activation energy," where any chemical reaction faced a barrier and, like an old man emerging from a chair, it needed a specific amount of energy before rising and crossing the room. It was Planck's plunge into electrolyte science that presumably brought him around to a belief in atoms ahead of most other physicists. By 1890, Planck told a colleague that physicists had "no alternative" but to embrace atoms and "descend into the molecular world."[23] The majority of physicists would take another 10 to 15 years to join Planck (and the frustrated Boltzmann) in this view.

He delivered an initial talk to his prestigious new colleagues at the German Physical Society's fall 1889 meeting. He stepped through his derivations, with precise chalk work, and declared his own significant contribution. But the room was silent. The chair of the department finally spoke, but only then with criticism. "I returned home somewhat depressed," Max wrote later, "but soon consoled myself with the thought that a good theory will succeed without clever propaganda."[24] Over time, Max would revise that maxim further.

"A new scientific truth does not triumph by convincing its opponents and making them see the light," he wrote, "but rather because its opponents eventually die and a new generation grows up that is familiar with it."[25] In some circles, this is known now as "Planck's Principle." Whether he intended it as a darkly humorous exaggeration or serious advice (most probably the former), the irony of Planck's Principle was his own lifelong flexibility.[26] Whenever evidence nudged him in a new direction, he could pivot. Just like he changed his mind on the existence of atoms, he later changed his mind on the (non)existence of the ether and championed Einstein's relativity. He even accepted, with great reluctance, some vexing new notions from 1920s quantum theory, after his intellectual baby had grown from an awkward teen to an enigmatic adult. (If we personify quantum theory, against all reasonable advice, the adult version wears the darkest sunglasses, never gives us a hint of facial expression, and sits in the corner texting to unknown recipients.) In the end, we will see that Planck was only reluctant to change his mind when it came to his own 1900 breakthrough.

Max rebounded from that first talk and quickly became a pillar of steady leadership and calm counsel for Berlin physicists. He would reach his happiest years with Marie in Berlin. They enjoyed four healthy children, their new villa in Grunewald, and evenings of music. They strolled through the local forest with family and friends. Their children played with those of other Berlin faculty, like Adolf von Harnack and the historian Hans Delbrück, up and down Wangenheimstrasse. But the Plancks had a regular retreat as well. Not long after moving into their new home, Max wrote an October 1905 entry for the letter diary. "My family was, as usual, at my in-laws in Tegernsee."[27] The Merck family's retreat in the German Alps, near the lake Tegernsee, became a second home for the Planck children. For many summers, Max would spend the first two weeks there, in the mountains south of Munich, acclimating to the altitude before starting his more serious tour of higher peaks (Figure 7.3).[28] "Already I feel in a few

Figure 7.3. Max Planck circa 1910, enjoying mountain air and something rare for him, a "blank mental state."
Courtesy Archiv der Max-Planck-Gesellschaft, Berlin-Dahlem.

days . . . reborn. There's nothing like country air, land, life, and the blank mental slate."[29] Other times, Marie would take the children to Tegernsee and leave Max to his work in Berlin, "a bachelor but for the letters."[30] He enjoyed writing to them, as they sat with her family and friends by the lake.

In the same era, Planck welcomed Lise Meitner into their home and into friendship with twins Emma and Grete. He threw himself into Einstein's fantastic world then, while expanding his influence as a scientist, writer, and editor. The appeal and influence of physics grew along with his own popularity; from 1889 to 1909, the enrollments increased nearly 10-fold in his lectures.[31] He wrote to Einstein in the summer of 1907 that they might be able to, with one sweeping theory, "unify all the forces of nature."[32]

But Marie was often ill. Despite their move to the cleaner air of Grunewald and the frequent doses of Tegernsee, she became increasingly frail. As Planck wrote to the letter diary in March of 1909, Marie began, "with a little cough, we placed no importance to the matter in the beginning, but gradually a fever came to her, and since the end of November to the present day she is confined to bed, with pneumonia on the left side." He took his wife to a sanatorium in Baden-Baden, "where she now hopefully finally goes to find healing."[33] However, by September, she had retreated one last time to her family's lake house. "My wife is not doing better," Planck wrote. "The fever remains constant at the same level—every night 38.5 [just over 101°F]—rising, of course, is unthinkable.... Currently, my wife is here in her old beloved room of her parents' house, and remains until the start of the bad season, around mid-October. Then I'll take her to Berlin and I hope that a tuberculin treatment can be attempted there." One imagines Max spooning her beef broth, as served to the sick in his Prussian youth. He closed, as usual, on an upbeat and hopeful note, telling his friends that "she feels particularly well lying in the open air on sunny days."[34]

Marie Planck never made it back to Berlin, and she passed on October 17. As Planck wrote to his cousin after the funeral, she was buried at Tegernsee, "the grave in which my lost happiness is sleeping. The daughter, next to her parents, as they take her back from the present.... Now, I'm facing the task to start a new life with my children. How this will happen is hard to see."[35]

For the year leading up to Marie's death, Max had been unusually brittle, and one imagines his wife's decline abrading his normal veneer. He had expressed exasperation when the all-but-certain Nobel Prize did not arrive in the fall of 1908. "[This] has so far brought me only irritation because of completely nonsensical newspaper reports," he wrote to the physicist Wilhelm (Willy) Wien, a close friend. "I ask therefore only for your silent sympathy."[36] He also lost his cool in the testy and nearly unprofessional 1909 public exchange with the elderly philosopher-scientist Ernst Mach, the appointed leader of the positivists, those who mocked the idea of fundamental truths. And just weeks before Marie's death, Max Planck challenged Einstein directly at a Salzburg conference when the pen pals should have relished their first in-person greeting. Einstein delivered a keynote address that surprised most of his audience, since he avoided any mention of his own relativity theory. Instead, he proposed that Planck's own work had irreversibly (if unintentionally) shredded all existing theories of light. This notion (correct yet rejected by physicists at the time) greatly upset Planck, so he stood up and gave Einstein a blunt piece of his mind about the young man's wayward ideas.[37]

After Marie's funeral, the children rallied to their Father. As he wrote to Wien, "I am starting so slowly to get used to regular life again, but it's hard. The children and work: these are the two sources from which I can get solace. It will get better with time."[38] At his next birthday, number 52, they brought in a cake and circled round their father, singing *Ein feste Burg* in four-part harmony.[39] But he would soon disappoint them.

Nearly a year after Marie's passing, he vacationed at Tegernsee with his daughters and spent time there with Marga von Hoesslin, Marie's niece. (He had actually known Marga since at least 1905, having mentioned, to the letter diary, her company on one of his alpine hikes.)[40] And in November of 1910, he shocked his children with an engagement: He would take Marga as his second wife.[41] Marga, 28, was closer in age to Erwin than to his father, whom she called "Uncle Max" into his later years.

Christmas that year was reportedly a quiet and awkward affair, as the children stewed and tried to understand their father's rush. Erwin, a steady journal-keeper, neglected to record an entry from this family holiday.[42] If the Planck home had seen its Camelot days, those departed with Marie. And through no fault of Marga's, the Planck family's long period of heartbreak could now begin.

8

April 1944

Max had two dear granddaughters—one by each of his twin daughters—and he helped raise them from infancy. Grete Marie (daughter of Grete) and Emmerle (daughter of Emma), both grew up in Germany's lean, chaotic interwar years. When Hitler came to power in 1933, Grete was 15 and Emmerle was 13. During World War II, Emmerle apprenticed as a nurse-in-training in the small city of Erfurt. She assisted with the floods of wounded soldiers and the growing ranks of civilians maimed or burned by Allied bombing.

In mid-spring of 1944, Emmerle climbed to a third-story window in Erfurt and threw herself into the air.[1] We do not know what placed her, at age 24, on that emotional edge. Max, the grandfather, blamed it on the "unstable" nature of Emmerle's father, Planck's son-in-law Ferdinand Fehling.[2] The ever-growing ranks of wounded, and the certainty of defeat, could not have helped. She survived the three-story fall but cracked a number of vertebrae, and she recovered in a local hospital, receiving care she'd sought to deliver.

Such attempts were increasingly common in Germany and culminated in a wave of mass suicides at the war's end.[3] The sad trend began after World War I, and despite Nazi propaganda claiming to have solved the problem, suicides continued at a disturbing clip from 1933 to 1939, when official statistics become alternately unavailable or unreliable. Some suicides were driven by fear of the Reich, especially when citizens were accused of racial impurity, homosexuality, refusal to work, or immoral behavior. As German war fortunes stumbled in 1942 and as the Allies began heavy bombardment, depression and suicides naturally increased. In fact, Emmerle's attempt occurred within a string that followed the massive bombing of Frankfurt that March.[4]

Despite his own increasing physical anguish, Max and Marga went immediately to Emmerle's side during her recuperation. Marga figures in Planck's tragedy as a fixture of bedrock in the personal quicksand surrounding them. She married her uncle Max and provided steadfast support throughout the most trying years of his life. She greatly enjoyed entertaining, supporting the *Geheimrat* (i.e., most respected, or "esteemed") Professor Planck's demanding social calendar and facilitating his work. She even accompanied him step for step on some of his challenging alpine climbs. The two of them could put a 20-something's spry youth to shame, as reported by their neighbor's son Axel von Harnack. He recalled one alpine hike from the years of World War I.

> During the ascent Planck did not speak a word nor did he rest even for a moment. He, a man of almost sixty, was decidedly superior to me, the young student, in his endurance and toughness. The first brief break was made after four hours ascent. We took a very modest war meal, then went on over four further mountain tops which made significant claims on our strength. A very brief rest on the summit, and down we went again. When we came home in the late afternoon, the Planck couple was surprisingly fresh, while I was at the end of my tether. Through-out the march Planck spoke very little, but his features revealed his great joy about the wonderful scenery around us.[5]

Max had known Marga for many years before their marriage, perhaps even in her childhood. She was born in Munich and would have been five years old—a ring-bearer's age—if she attended Max and Marie's wedding there. The decision to remarry has the whiff of Max Planck's pragmatism: He maintained his ties to the Merck/von Hoesslin family, including fond trips to see them in the German Alps; he quickly filled the critical vacancy of faculty wife; he recruited a hostess and mother figure for the household; and he secured a younger partner, both hale and willing as he approached the later stages of life (Figure 8.1). After his death, she wrote that it was an honor to have lived with and assisted such an august man.[6]

After Marga joined the household, she conceived at once, and on Christmas Eve, 1911, she gave birth to Hermann Wilhelm Heinrich Planck, their only child together. Max was overjoyed, and the family took to calling Hermann "Bubie." It was a special delight, for instance, when Bubie, presumably in Marga's arms, brought his father a daisy on April 23, 1912, Max's fifty-fourth birthday.[7]

To take stock, we pause in 1912. Max had lost Marie, but he was by no means lonely. He had already married again and found himself

Figure 8.1. Marga and Max Planck in Grunewald, 1924.
Courtesy Archiv der Max-Planck-Gesellschaft, Berlin-Dahlem.

surrounded by five healthy children (spanning Hermann in infancy to Karl at age 24). He also had two other remarkable women in his life: his mother Emma and Lise Meitner. His discovery of 1900 was, at long last, attracting more attention and discussion; he had just bathed in the stimulation of the first scientific meeting geared to discuss the new quantum theory—the first Solvay conference in Brussels—and he was but one year from seeing the Danish physicist Niels Bohr take the original quantum hypothesis and revolutionize the very notion of matter itself. In 1912, the Prussian Academy of Sciences elected Max (unanimously but for one anonymous vote), to the prestigious Secretary position. His neighbor Adolf von Harnack had spearheaded the formation of a new research organization, the Kaiser Wilhelm Society, devised to enable grand scientific work that was too expensive for individual universities. And Planck was already scheming to bring Albert Einstein to Berlin, placing the crown jewel atop German science and securing its international standing for, he must have assumed, decades to come.

But worries simmered within the family. The depression that would push his granddaughter Emmerle from a window in 1944 must have been familiar to Planck. Max's daughter Grete had been briefly institutionalized with a "nervous attack" at an eerily similar age, around her twenty-fourth birthday in 1913. After two weeks in a sanatorium, she reportedly emerged with changed perspective and a focused life plan.[8]

But the most serious case of emotional torment played across the brief life of Karl Planck, Max and Marie's eldest child. As a young man adrift with no particular drive or discipline, he faced great tension with his father, a man with nearly inhuman focus. "My eldest, Karl, who was originally becoming a lawyer," Planck relayed to the letter diary in 1908, "is changing seats and has become a geographer—a worrisome jump." Max tried to talk some sense into his son, but it was "of course, not pleasant."[9] When Karl lost his interest in geography and sought to switch his focus to art history, Erwin stood up for his older brother against their doubtful father, to no avail. The art-sympathetic Karl was a piece of whimsical kinetic sculpture compared with Max, who worked like one of the humming machines driving Germany's industrial economy.

Karl soon lost his connection to sleep, and he struggled to make himself eat. "My brother Karl has had a nervous breakdown," Erwin wrote to a friend in early 1912. Max committed Karl to a mental hospital in Kassel, again at an onset age of 24. Erwin went to stay with Karl over an Easter vacation, and the family soon saw improvement. But after leaving in May, Karl continued to struggle. The arguments with his father rose like regular thunderclouds, casting shadows across Planck family gatherings. In early 1913, Max wrote to Erwin that, "Karl gives me great grief again."[10]

As Max delivered his first talk as the University of Berlin's rector, in the fall of 1913, he cautioned against the lack of focus he saw in Germany's youth. He held up the energy of America as a worthy example to them. This was nearly a personal comment, because in those days Karl still struggled. He shuffled from one job to the next without success, short-circuited by a nervous depression.[11]

The war interrupted Planck's years of worry for Karl in 1914, and it appeared to many Germans as a blessed solution for all the empire's ailments. The war brought purpose. As Erwin headed to the Western front as an officer, Karl enrolled in artillery school, and the twins volunteered as nursing assistants in various hospitals.[12]

The entire previous year had seen a fever pitch of nationalism as Germany loudly celebrated the 100th anniversary of chasing Napoleon Bonaparte back to France; the Empire celebrated its rising power and ambitions as well. Unlike most Germans (even German academics), Max kept a very calm tone throughout the frothing of 1913. As a newly appointed rector, many expected Planck to sound more patriotic. As festivities began in 1913, Planck noted that the professors of 1813 had, "proved their patriotism in their own way, through calm and true fulfillment of duty, just as much as the young soldiers who fought in the field for the liberation of the fatherland."[13]

When the war arrived to a giddy continent, he delivered his second rector's address to the university on August 1, 1914. He knew that his own sons and those of his colleagues would be headed to battle. He knew his lecture seats and the university's hallways would soon be emptied of students. His remarks were incredibly prescient.

"We do not know what tomorrow will bring; we only suspect that something great, along with something monstrous, will soon confront our people, that it will touch the life and property, the honor and perhaps the existence of the nation." Planck's standard modus operandi in speaking and writing involved painting oppositions—in reading Planck, one starts to anticipate the next "however." In this address, it was much more than a tic or habit. If Friedrich Nietzsche had given Germany the notion of "will to power," Max explored a will to optimism. He felt himself pulled to the fervor of the moment as he continued that 1914 address. "But we also see and feel how, in the fearful seriousness of the situation, everything that the country could call its own in physical and moral power came together with the speed of lightning and ignited a flame of holy wrath blazing to the heavens, while so much that had been considered important and desirable fell to the side, unnoticed, as worthless frippery."[14] And he told his students, as they prepared for war, "Germany has drawn its sword against the breeding ground of insidious perfidy."[15] No one complained about a lack of patriotism this time.

He echoed these sentiments in letters. In the following month, Planck wrote to a relative that, "it is a great feeling to be able to call oneself German." He particularly craved the unity that war brought, and he welcomed stripping away the trivial concerns and politics with which people normally wasted their days. As he wrote to friend Willy Wien a few months later, despite the admitted horrors of the conflict, "there is also much that is unexpectedly great and beautiful: the smooth solution of the most difficult domestic political questions by the unification of all parties."[16]

In a book that greatly exceeds its own provocative title, Peter Englund's *The Beauty and the Sorrow* presents a series of journal entries from those living in the midst of World War I. The exaltation resonating throughout the German Empire is difficult to understand with just a century's remove. A German schoolgirl reports her headmaster overcome by tears of joy with the announcement of war. The schools quickly banned common non-German words, such as "mama" and "adieu." And with each announcement of a battle victory, the children were allowed to scream in class, the louder and longer the better.

Train stations all over the Empire hosted celebrations. Row upon row of marching grey uniforms synced pounding boots and beating drums. Marching bands would strike up every crowd's favorite, *Die Wacht am Rein*, the Watch on the Rhine, and the assembled would sing at full throat.

Dear fatherland, put your mind at rest.
Fast stands, and true, the Watch, the Watch on the Rhine.

Happy soldiers waved from train cars before departing, with "garlands of flowers around their necks or pinned on their breasts. Asters, stocks and roses stuck out of the rifle barrels."[17] Promises of soldiers could be heard over the steam engines: We'll be home before Christmas. The journal of one French soldier encapsulates the feeling of so many then, whatever their nationality. "How humiliating it would be not to get to experience the greatest adventure of my generation!"[18] This spirit swept Karl and Erwin Planck smiling into uniform, with a proud father looking on.

The overwhelming majority of German intellectuals, Planck included, stepped forward in mind, heart, and rhetoric. Most scientists agreed with what Adolf von Harnack had told Kaiser Wilhelm: "Military power and science are the twin pillars of Germany's greatness."[19] Albert Einstein however, thought himself surrounded by madness. After Planck's aggressive recruiting, Einstein had just started his new post in Berlin in 1913, only to be reminded of the rampant militarism of his Munich childhood. He subscribed to a politics much more liberal than Planck's, aligning himself with the ultraprogressive New German League, a party banned by the Kaiser's government. Weeks after the outbreak of hostilities, Einstein wrote to his friend Paul Ehrenfest that mankind was "a sorry species" if it could celebrate war.[20] Later he wrote, "that a man can take pleasure in marching in fours to the strains of a band is enough to make me despise him . . . [and he] has only been given his big brain by mistake; unprotected spinal marrow was all he needed."[21]

Lise Meitner's view was more typical of academics at the time, as she supported the war, particularly as a native Austrian. She wrote to her colleague Otto Hahn about a night of entertainment and conversation at the Plancks in 1916. "They played two gorgeous trios, Schubert and Beethoven. Einstein played violin and on the other side volunteered some naïve and strange opinions about politics and war. Just the fact that there is an educated person around who in this time does not touch any newspapers is certainly curious."[22] And this was presumably Planck's view of Einstein as well: The wacky politics were just a curious side effect of such genius, drifting in thoughts of outer space (and time).

Another of Planck's colleagues frames the other end of the academic spectrum. The great chemist Walther Nernst lost two sons in the first 18 months of the war, but proudly so. He volunteered to drive an ambulance at the front, and he could be spotted, in uniform, practicing his high-step in the front lawn while asking his wife to critique his form.[23] This was a not-so-unusual example of delirious commitment to World War I in its first stages.

The singing in the streets of Munich escorted Max's mother, Emma, from life on August 4, shortly after his rector's address in Berlin. His lifelong close friend died at 93, vigorous and lively into her later months. He was unable to attend her funeral in Munich because war mobilization trumped civilian rail travel.[24]

Despite losing the two women dearest to him within five years, he remained bullish on the future of the family, the empire, and physics. His sons pursued honor, the Kaiser was righteous, and German physics promised years of dominant progress with Einstein at the lead. One wonders if Planck ever regarded his own sorrow at losing mother Emma and wife Marie as "frippery" when compared with the great uniting mission confronting the empire.

The war started with a steady stream of German victories. While most Germans assumed Russia was the primary enemy and that all trains should be hauling soldiers to the east, military advisors had persuaded the Kaiser to quickly neuter France, a primary Russian ally. And given Germany's last success against France, in 1870, most military leaders thought it a sane strategy. As some crowds looked on in confusion, a number of military trains headed west. With Erwin Planck in their midst, the first German units marched quickly through Belgium and into France, pushing French units into full retreat. This new conflict began to look as easy as the Franco-Prussian War, but the success only lasted a few weeks.

The more sober reality hit home by October, when the first battle of Ypres killed 40,000 German soldiers in 20 days—at home, citizens called it the *Kindermord*, or "the children's slaughter." In a single battle, Germany suffered twice the losses it had in the entire Franco-Prussian War. A German schoolgirl wrote in her diary about helping the Red Cross feed departing and returning soldiers. On a biting December night, with snowflakes drifting through the light of gas lamps, they made hundreds of onion sausage sandwiches and carried vats of pea soup to the troops. She noted the sharp difference in the trains. For everyone departing with eager and singing soldiers, another would come back, full of silent and damaged men.[25]

What would Erwin or Karl have experienced on their way to battle? If they were among the lucky, they shipped in a train with windows, though many German soldiers moved hundreds of miles shaking to and fro in darkened cattle cars. Transit could take as long as four days. Detraining, they would walk the last miles toward the front. Soldiers reported seeing the orange muzzle flashes from a great distance, looking nervously at the jagged, angry horizon, and then approaching an ever-louder realm, drained of color and shredded by artillery. One young soldier deployed near Liege wrote in his journal, "Where there had once been tall white houses with shutters on their windows, nothing remains but spiky, rain-blackened heaps of rubble, bricks, and splintered wood. The projectiles from shrapnel shells and shell fragments lie scattered all over the streets. The little town is slowly being ground down into the earth."[26]

By Christmas of 1914, the barrage of bad news had choked all hopes of a speedy conflict, and the extent of Europe's mass delusion became clear. Gone were dreams of daring exploits, crushed by a new reality. The chemistry of explosives and the engineering of firearms made humans almost irrelevant. Standing men were clutter to be swept to the side in violent bursts. There was no longer a good soldier or a weak soldier—just lucky or unlucky ones. As of December of 1914, the surreal networks of trenches were already set and spreading rapidly as long stretches of barbed wire unfurled across the ravaged landscapes. Though frequently cold, muddy, and miserable, German soldiers and officers helped themselves to French interiors. "Shelters in the trenches are cheaply and gaudily furnished with loot from French homes," wrote the same soldier near Liege. "Everything from woodstoves and soft beds to household equipment and beautiful sofas and chairs." Eventually, some trenches put up electric lighting, and even wainscoting.[27]

Karl Planck would have seen these incongruous details from a distance: Such comforts were typically reserved for officers. Karl would have turned to his own gear for respite. A German soldier's standard kit list from 1914 included, for example: one aluminum mug, gray gloves, two tins of coffee, one tin of rifle grease, two bandages, two pairs of underwear, four pairs of socks, knee warmers, a white armband for night fighting, one speckwurst sausage, the New Testament, 30 field postcards, and anise oil (as an antiseptic).[28] Though the standard kit also included 150 rounds of ammunition, most soldiers like Karl fired very few shots.

Life in the trenches involved a lot of boredom. Days or weeks could pass in relative silence, with the men writing postcards, digging trenches into the night, or eventually playing pranks on one another, such as putting pepper in one another's gas masks, or "pig snouts," as the German soldiers called them. Gunfire and shelling could start up at random, or from something as senseless as a soldier shooting at a bird. Several dead bodies later, silence would return with just as little explanation.

German soldiers reported the hunger of French citizens around them. One wrote of the horrible waste perpetrated, as soldiers spread oats on cobblestones to muffle the movement of artillery pieces, all while hungry civilians looked on.[29] The most severe food shortages were in Germany and Austria. Conscription of farmers and even farm horses into the military greatly hurt domestic food production, and the British dominance of the northern seas effectively cut Germany off from food imports. As early as January 1915, newspapers carried warnings like the following: "Any individual using corn as animal fodder is committing a sin against the Fatherland and may be punished." By 1917, the average German had lost 20% of his or her prewar weight.[30]

As of March 1915, two of Max Planck's nephews had been killed in the fighting, and Carl Runge's son Bernhard died in the *Kindermord* in 1914.[31] "Where is the compensation for all this unspeakable suffering?" Planck wrote to the beloved Dutch physicist Hendrik Lorentz.[32] And Max suffered daily on account of Erwin.

In September of 2014, Erwin went missing at the battle of the Marne, Germany's first setback. After miserable weeks looking for answers, Max and Marga learned that he was wounded but alive as a prisoner in France. In time, they were allowed to exchange letters, as long as they wrote in Latin. "Thank god he at least is healthy," Planck wrote to his cousin's family. "And has been writing with good courage in spite of bad treatment he's receiving at St. Angeau."[33] The empire bestowed a number of medals on Erwin, including one recognizing battle wounds, and Max collected all of them.

Erwin Planck had pursued officer's training well before the war, with financial support from his father. When he earned his *Leutnantspatent* in August of 1912, Max learned of it in the newspaper and sent his son a proud telegram.[34] Erwin completed officer's training in the spring of 1913, joined the military reserves, and immediately began medical school. After just a year, the war called him to duty, where his officer's status presumably helped preserve his life as a prisoner. In the prison camp, as in the trenches, rumors constantly percolated. The war would end next year since everyone was running out of bullets. Montenegro surrendered to the Austro-Hungarian Empire, so surely the war would end within months. *Et cetera.*

After they established regular communication with Erwin, including Max reading his letters aloud to Marga and daughter Emma, the rest of the Planck clan weathered the war's first months in relatively good spirits. Daughter Grete had studied violin within Heidelberg's music conservatory where she met and, in the war's first weeks, married a middle-aged history professor named Ferdinand Fehling. Max told his friends that Grete was "war married," as Professor Fehling joined the war effort and deployed to the northern border. Planck had not met this fellow, but at least he was of Prussian stock, born and raised in Lübeck, near Kiel.[35] Given her recent struggles with depression, one can imagine Max Planck's relief for his daughter, though he would have preferred a more traditional courtship (Figure 8.2). Emma went to meet her new brother-in-law and reported to Max that they would, "get along well . . . he is a very vivacious, spirited, and artistic man."[36]

Emma herself enjoyed her early days of hospital work, referring to the wounded soldiers as "her little sheep." She wrote to Erwin regularly, saying she missed her siblings terribly but said it was, "wonderful for me to be doing a job that is important during the war." She led her wounded flock

Figure 8.2. Grete (left) and Emma (right) Planck in 1906.
Courtesy Archiv der Max-Planck-Gesellschaft, Berlin-Dahlem.

in singing together, and for those on the mend, even took them outside for sledding and snowball fights in the war's first winter.[37]

Against all odds, worrisome Karl was thriving. His father later wrote that his son was "one of those healed by the war."[38] Emma and Max relayed to captured Erwin that Karl was better than ever, achieving a promotion to Sergeant in the summer of 1915. After his first battle wound, catching shrapnel in his shoulder, he visited the family home in November but yearned to return to the front.[39] Most wounded who could still walk, talk, and use their arms had to return to the fight, and Karl was no exception.

In February 1916, the German army launched a new offensive, trying to break the will of the French near Verdun. Over the next 10 months, the intensive fighting would yield no more than 10 kilometers of progress to repay avalanches of casualties.

We can see Verdun through the diary entries of the Frenchman Rene Arnaud. As a leader of his unit, he was told, "the whole thing is very simple. You will be relieved when three-quarters of your men have been knocked out. That's the going rate."[40] And so it was for both armies at Verdun, where the two sides matched one another casualty for casualty. In the end, the nearly three quarters of a million casualties ranks Verdun as one of the most deadly single battles in human history. If we printed the names one per line, single-spaced, we would have a 15,000-page book with a binding six-feet thick. Severely wounded in May, Karl Planck drifted from life amid this heap.

Planck wrote to his cousin's family that they could not confirm his death "with absolute certainty" but knew it must have been true given "intense" injuries. "My dearest wish is that he didn't suffer any more. The pain of this war you only feel for real when you feel it in your own flesh and blood. There were times where I was not without sorrow for Karl's future. Then his life didn't seem so precious to me as it's been presented to me now." Ever mindful of his audience and of his cousin's own concerns, he returns to patriotism, despite the absurdity of the battle, the meaninglessness of another useless, pulverized fort, and its human price. "We did not know that Fritz is stationed near Verdun. I hope that he will help seize it. This would be a great step forward."[41]

Before the war, Max had prepared himself for years of suffering with Karl, as his troubles appeared beyond resolution. The war provided a perverse hope for a life of meaning and then it severed any long-term concern. "Without the war I never would have known his value," Planck later wrote to Willy Wien. "And now that I know it, I must lose him."[42]

Meanwhile, Planck was busy keeping German science inching forward with less funding and fewer pairs of hands for laboratories and notebooks. In 1916, the Kaiser himself appointed Max as senator to the Kaiser Wilhelm Society, overseeing and promoting the empire's scientific enterprise. Planck's friend and colleague Einstein deeply imbedded himself in what he called the greatest idea of his life. Against Planck's advice, he spent most of the war year's refining a bold rewrite of Newton's gravity. And Lise Meitner threw herself further into radioactivity research, despite Otto Hahn being called to serve in the war. Before the war, she had already expressed misgivings about her tunnel vision, and these echo the feelings of many scientists, both in and out of wartime. As she wrote to a friend, "what distresses me most is the frightful egotism of my current way of life.

Everything I do benefits only me, my ambition and my pleasure in scientific work. It seems I have chosen a path which flies in the face of my most deeply held principle, that everyone should be there for others."[43] Although Meitner penned this shortly after her father's death, such sentiments could not have waned during the war's devastation.

By the end of 1916, the Central Powers offered peace to the Entente, but the overture was quickly rejected. Though not yet at war, the American president Woodrow Wilson spoke with a firm voice against the offer, saying that without more specifics, such overtures were meaningless.[44] The war lurched forward. As of Christmas 1916, Planck had lost his oldest child, while another child concluded his second year in a prison camp. The only sparkle of positive news came from Grete: She was pregnant, headed for a springtime birth and a new generation. While it was a difficult time to start a family, Planck the optimist welcomed the news.

But an early 1916 frost cut the German potato crop in half, and the German Empire entered its infamous "turnip winter." With the British naval blockade squeezing the German food supply harder than ever, the nation's overreliance on homegrown potatoes left no room for error. Even before the war, the average German ate nearly 1,300 pounds of potatoes per year.[45] The failure of the potato crops in 1916 left a disastrous void in German pots and stomachs. Turnips provided the only available substitute, and they became known as "Prussian Pineapples." Families faced mashed turnips, turnip soup, and turnip salad. Those seeking variety tried turnip pudding and turnip balls with turnip jam.[46] Disease flourished among the malnourished and ever-hungry civilian population. Planck himself, famously never ill, was bedridden in December of 1916 with a bronchitis bordering on pneumonia.[47]

In 1917, approaching a thirty-eighth birthday, Einstein developed a debilitating stomach ailment. Poor nutrition exacerbated or even caused his suffering, but his personal life couldn't have helped. Separated since his move to Berlin, he and his wife Mileva Maric now discussed divorce. He once wondered aloud to her which would conclude first, the war or their divorce process, and in fact, the armistice of World War I preceded the divorce by a number of months.[48] Whatever the causes, he lost a dramatic 50 pounds in 1917 and a great deal of energy in the bargain. And so began his famous penumbra of graying hair.

The desolate nutrition also hurt first-time mothers like Grete Planck. Germany suffered an infant mortality rate of about 19% in the turnip winter.[49] Mortality among women also spiked that year, rising 30% above its prewar levels.[50] Grete gave birth to a healthy baby girl in May, but then a blood clot docked in her lungs, causing a major embolism. One moment, she would have been happily nursing her one-week-old baby, Grete Marie Fehling, and then the quiet would burst with sudden symptoms: elusive breaths, a violent new cough, and chest pains. She died there in Berlin, with Max Planck nearby, and the family placed her ashes in the Grunewald cemetery. Emma informed Erwin with a heart-wrenching letter. "Especially I'm thinking of you, my poor dear muse as you are far away and must now endure a second enormous anguish," she wrote. "I cannot imagine that our good dear sister, my inseparable twin, should be here no longer.... Grete's life had only started now and how she would have made everyone else happy with her joy, especially those who knew her so well and went through hard times with her." She writes of a photo of all four of them, how they naturally belonged together. "Let's stick together, my dear muse, more tightly than ever." And she closes with sympathy for what their father has endured, losing two of them. Max also wrote to Erwin, saying, "I wish I could be by your side ... to survive this blow together and support one another." But he urged Erwin to trust in better times to come.[51]

Max would assume a major role in Grete Marie's upbringing, and naturally, Aunt Emma stepped forward to help Professor Fehling as well. Emma was a striking replica of her departed sister. Anytime Emma entered his home, Fehling must have started as if visited by a beautiful ghost.

When Erwin gained his freedom and returned home in October, still hobbled by a war-wounded thigh, he found sister Emma waiting for him at the train station. And at Wangenheimstrasse, he found a profoundly changed family, with two siblings departed and a new motherless niece. He spent time with his father and sister playing music, and they traded readings from their respective journals of the last three years.[52]

Incredibly, Planck found a way to stay upbeat. His longtime dream of a Kaiser Wilhelm Institute for physics came into being in 1917, though it existed more in letterhead than fact, given its lack of a physical home. In a December letter to Einstein—the appointed director of the new institute—he advised his friend on a number of administrative matters and then included a cheery personal note.

> May the New Year bring you full recovery from your health complaints! I hope that during the same year your sympathies for the German Party grow

as well; it is ever ready for peace, despite our being in such a favorable position militarily now as never before. I am also pleased with the rapid rise in our exchange rate.

So, on the 2nd we have colloquium. Will the room be heated then, I wonder?[53]

What could cause this optimism, especially as America and her resources had joined the Entente in 1917? The Bolsheviks had assumed control of Russia in November, and negotiations of a ceasefire had begun in early December. The eastern front would close, leaving more resources for the main fight to the west. And the Central Powers had made progress that autumn against Italy. Indeed, a new German offensive throughout the spring of 1918 brought them to within 60 miles of Paris. (Such bursts of positive war news set Germans up for shock and disbelief at the war's resolution.)

In 1917, the German potato crop rebounded, but overall scarcity of basic foods and supplies strangled the Central powers. In 1918, babies received boiled rice instead of milk. Peter Englund writes of fake meats made of pressed rice, stabbed with a shaft of wood, carved to resemble a bone. The Kaiser's government recognized 837 substitutes for meat, and 511 for coffee.[54]

Given severe rationing and empty stomachs, the Berlin science community welcomed the excuse of celebrating Planck's 60th birthday with a banquet that spring. Planck's friend Einstein stepped forward as master of ceremonies, but he worried about his performance. "I'll be happy tonight if the gods grant me the gift to speak profoundly," he wrote to Paul Ehrenfest, "because I am very fond of Planck. And he will certainly be very pleased when he sees how much we all care for him and how highly we value his life's work."[55] Indeed, affection and respect rained on Planck at this celebration. The general (but not unanimous, as we will see), sentiments for Planck were encapsulated by the words of physicist Max Born. "You can certainly be of a different opinion from Planck's," he once wrote to Einstein, "but you can only doubt his upright, honorable character if you have none yourself."[56]

By July, Allied counteroffensives relentlessly pushed the German army back. By early November, the Austro-Hungarian dual monarchy was no more, and they signed an armistice with the Allies. Meanwhile, sailors in Germany's high seas fleet were beyond the limits of their patience. In Kiel, they mutinied against officers' orders to return to their ships. A general strike throughout Germany followed, and Kaiser Wilhelm II abdicated his throne on November 9, with an official end to conflict two days later. The exhausted and malnourished citizens of Europe had little time to celebrate the war's ending as a devastating flu pandemic swept the continent that winter, taking another two and a half million lives, or one in every hundred remaining people.[57]

Postwar Berlin fell to chaos. Lise wrote with worry for her family in Vienna, but a friend wrote back: "I am almost more concerned now about Berlin than Vienna."[58] Planck wrote to the letter diary that Christmas of, "the final defeat, and even worse, the inner struggle, in which the remnant forces tear at one another."[59]

Rival political factions exchanged gunfire across the Spree River, bordering the University of Berlin. A group of socialist students occupied university buildings and seized the rector (not Planck at this point), along with several deans. Einstein was called to mediate because he was the one university voice that the leftist students trusted. After several tense conversations, the students relented and freed their captives. With his colleague Max Born, Einstein walked directly from the Chancellor's office to a nearby gathering at the Reichstag, and he delivered some prepared remarks warning against the possible excesses of the far left. He called for immediate elections, to eliminate fears of a new form of dictatorship. Twenty-five years later, with Hitler in power, he looked back on this day with dismay. "Do you still remember the occasion?" he wrote to Born. "We were so naïve for men of forty."[60]

Most of the physicists were less inclined to make political statements. They decided to reinstitute their colloquium series (weekly talks for hearing about the latest research findings), and Meitner described the herd of physicists ambling into the deathly cold lecture hall, most draped in their old army coats, with gunfire echoing in the streets outside.[61]

In the first half of 1919, the treaty of Versailles emerged, placing the punitive weight of the war squarely on Germany's back. It aimed to ensure that the Empire would never again threaten its neighbors. As the economist John Maynard Keynes wrote at the time, "the economic consequences of peace" would be catastrophic for both Germany and all of Europe.[62] Most Germans, including academics, were dumbfounded by the conditions. Planck's colleague and Nobel laureate Emil Fischer (who had once feared allowing Lise Meitner into a chemistry laboratory), lost two sons during the war. In May 1919, he wrote to the great Swedish chemist Arrhenius that, "we are all shocked by the conditions of the new treaty, which are so cruel to Germany."[63] His death in July, following a diagnosis of intestinal cancer, was reported as a suicide in some circles.[64]

The only tentative bright spot for Planck involved daughter Emma, who was married on February 1, 1919, and soon carrying another grandchild. The new groom and father-to-be made for a familiar son-in-law: Professor Fehling. He had fallen for the lively ghost of his first wife as they cared for little Grete Marie together.

By August 1919, a new constitution and republic congealed from an assembly in Weimar: A liberal democracy would try to lead its depressed and skeptical population, with every family reeling from war casualties, hunger, and influenza. The new government also faced staggering war debt, and the first signs of an accelerating inflation. Printing reams of money appeared to be the only way to pay their bills.

In the postwar months, Planck continued his ascent to the forefront of German science, as a spokesman and guide. He gradually assumed the position of his former teacher, Hermann von Helmholtz, as the public face of science not just in Berlin but Germany overall.[65] He presented a rare and optimistic public voice. "The main thing," he expressed to colleagues after the Versailles announcement, "is not to lose courage and the hope that better days will come." And in a Christmas day newspaper column, he told the public that German science would ensure that the Fatherland would remain "in the ranks of civilized nations," despite all signs to the contrary. He saw his passionate optimism as a "necessary prerequisite" to survival, and he urged others to envision a day when the Empire was even better than it had been in early 1914.[66]

His standing only improved in October when the Nobel committee in Stockholm announced that Max Planck had won the delayed 1918 Physics prize, for "the services he rendered to the advancement of Physics by his discovery of energy quanta." Simultaneously, the committee awarded the 1919 prize to another German physicist, Johannes Stark—a man increasingly adversarial to both Planck and Einstein—for his laboratory exploration of single atoms responding to electric fields. (The "Stark effect" describes how an electric source affects a nearby atom's internal structure.) Max took part of his $1.25 million reichsmarks prize and invested in Erwin.[67] His son now sought a career in diplomacy, and he was headed back to school. The investment was well-timed, because the sum would only have evaporated in the coming inflation, had he tried to save it.

Planck had little time to enjoy his Nobel Prize. In November, even as his second grandchild, Emmerle, entered the world, her mother, Emma, died from complications. It was as if nature repeated some cruel experiment with Professor Fehling and each of Planck's twin daughters. Neither had lived past the age of 30, and Max now fell fully into a depression. He buried Emma's ashes in Grunewald. "The two beloved children, who could not get along without one another in life," he wrote to relatives, "are now together for ever."[68]

"Planck's misfortune wrings my heart," Einstein wrote to a friend. "I could not hold back the tears when I saw him. . . . He was wonderfully courageous and erect, but you could see the grief eating away at him."[69] Planck wrote to his most trusted colleague in physics, Hendrik Lorentz, "There have been times when I doubted the value of life itself."[70]

Incredibly, he kept himself upright, throwing himself back into work and cherishing his remaining family: Marga, son Hermann, baby Grete Marie, infant Emmerle, and more than ever before, his son Erwin. The New Year brought back at least a glimmer of his deep optimism. "There are still many precious things on the earth and many high callings," he wrote to a relative. "And the value of life in the last analysis is determined by the way it is lived."[71]

Looking back, he would later credit his religious faith as the main source of strength in tragedy. While he did not subscribe to any specific Christian vision of God, he suffered no doubts in the existence of divine forces working in the universe. "If there is consolation anywhere it is in the Eternal," he wrote to a colleague just a year after adult Emmerle attempted suicide in Erfurt. "And I consider it a grace of Heaven that belief in the Eternal has been rooted deeply in me since childhood."[72]

In the weeks and months after Emma's passing, solace sometimes arrived at the Planck home carrying a violin case. The trio of Einstein, Erwin, and Max Planck mourned within their music.[73]

9

May 1944

Max and Marga left his granddaughter Emmerle recovering in Erfurt and took a long, multistop train ride to the tiny Bavarian town of Amorbach. But Max now suffered incredible pain radiating from his pelvic floor, and his arthritic back had never been worse. A doctor in Amorbach diagnosed Planck's severe hernia and said the 86-year-old would need surgery, the sooner the better. Max wrote to Erwin for advice. Erwin and Nelly quickly found and loaded a well-known surgeon in their car and drove the hundreds of miles to Amorbach.

Max was awake and alert throughout the operation, and because of substantial blood loss, Erwin provided a transfusion.[1] It would give a unique new level to their bond, with the literal last remnants of the mother passing from son to father. We can imagine Max attempting to joke about feeling younger and better looking by the drop and Nelly telling him to shush. If father and son locked eyes, did Erwin wonder about keeping secrets from his father? Erwin's letters of that spring had repeatedly told his father of his yearning for the war's end. During this urgent trip to Amorbach, did Erwin contemplate the coming attempt on Hitler's life?

After the operation, Erwin and Nelly drove back to Berlin, while Max and Marga eventually took a train to Rogätz. Marga wrote Max von Laue of the surgery's success. She said her husband had been a good and understanding patient, but she worried now about the signs of heart disease and the continued misery in Planck's spine. The vertebrae were beginning to fuse.[2]

Their trip through the middle of Germany would take them around or snaking through the ruins of Leipzig, which had been largely destroyed in a December 1943 bombing. The resulting firestorm was reportedly comparable to that in Hamburg, but there were fewer casualties, as Germans had learned to flee their shelters after the last bomb's blast and before the fires came for them. The author W. G. Sebald was a small child at the time, and he recalled associating the word "city" with heaps of rubble. Many cities like Cologne were reduced to two-story drifts of debris. Sebald described trains filled with refugees who refused to look on the devastation—anyone gazing through a window was known immediately to be a foreigner. The streams of homeless (truly townless) Germans carried "disquiet along little rivulets and into the most remote villages." Many of them stumbled about so dazed that drivers, like Erwin, had to constantly avoid running one down.[3]

Nazi leaders recognized what was coming. They were now in retreat. At the time of Planck's surgery, the Reich began moving heavy machinery from France to Germany. In July, they trucked massive deposits of gold out of the Fatherland into their accounts in Spain, Portugal, and Switzerland. In August, Heinrich Himmler drafted plans for keeping the Nazi movement alive in a postwar world.[4]

Despite the horrific losses, the Nazi government gripped power as securely as ever. Hitler explained this to his chief architect Albert Speer:

> Even the worst idiot realizes that his house will never be rebuilt unless we win. For that reason alone, we'll have no revolution this time. The rabble aren't going to have the chance to cover up their cowardice with a so-called revolution. I'll guarantee that! No city will be left in the enemy's hands until it's a heap of ruins![5]

Although Hitler had ceased public statements or appearances, Joseph Goebbels continued serving a precise mix of fear and hope. *Imagine what will happen if the Russian barbarians overrun the Fatherland*, on the one hand, versus *new devastating weapons will soon be introduced to our enemies*, on the other. As Nazi press officer Hans-George von Studnitz wrote in his 1944 diary, "Since no one has an overview of the situation or knows a way out of it, since everybody is afraid of making developments even worse by disloyal behavior, the regime can go on counting on the people's support. In many respects, the situation today is different than in 1918. The morale of the homeland has, despite the burdens inflicted by the air attacks, remained intact."[6]

As Erwin drove through the ruined Fatherland, he knew a group plotted against Hitler, but as far as we know, he had guarded the secret. And with Erwin employed in industry, removed from politics for all appearances, Max probably didn't worry about his son running afoul of the Gestapo. But Erwin hadn't ceased his political activity. In 1939, as a consultant to the High Command of the Armed Forces, he'd helped draft a memo urging Adolf Hitler to stay out of Poland, explaining why an invasion would lead to a larger war, and, in logical turn, the nation's ruin. (Hitler never received the memo, as an underling intervened.) And as recently as 1942, Erwin had attended several high-level resistance meetings concerning a potential coup.[7]

Max Planck presumably never knew about these, and he'd last truly worried about his son during the early days of the Weimar Republic. Erwin had suffered existentially in the early 1920s, a sensible state for any survivor of World War I and so much personal loss. After completing his diplomacy training, Erwin found work in the Defense ministry thanks to his wartime friend Kurt von Schleicher, an army general who became ever more powerful within the Weimar Republic.[8] In retrospect, von Schleicher's will to diminish democracy, fuse military and civilian life, and elevate the power of the presidency all helped realize the Nazi party's dominance. Erwin yoked himself to this military friend. His choice sheds light on his conservative politics and arguably Max Planck's as well.

In August of 1920, Max, Erwin, and Marga took a hiking trip into the Alps. Erwin and Max had a long history of talking politics, and this must have intensified when Erwin entered the new Weimar government. If they made their usual stop at Lake Tegernsee to visit the Merck and von Hoesslin families, they may have noted a strange scene. From a pier, a small group of fringe radicals held out a long pole and slowly unfurled a new, home-stitched flag. The red banner featured a lone, black *Hakenkreuz* (hooked cross) whirling within in a white circle. The group had just rebranded themselves as the National Socialist German Workers Party (NSDAP). By the time of this Tegernsee flag raising ceremony, their fifty-fifth member had risen to official spokesperson and director of propaganda. Adolf Hitler, a young veteran of the Bavarian army, began igniting crowds with speeches that year. But if Planck and his in-laws sat lakeside that day, they would most probably have just laughed at the kooky-looking group and turned back to their conversation.

Pervasive anti-Semitism leaked into daylight well beyond fringe political groups. These feelings had retreated in Germany since the later nineteenth century, but they were never absent, and two factors brought them gushing back. Public sentiment often blamed Jews for black market profiteering during the hungry years of World War I. At the same time, a significant wave of *Ostjuden*, or eastern Jews, flowed into Germany as they fled renewed persecution in Bolshevik Russia. Between 1910 and 1925, Berlin's Jewish population quadrupled.[9] Brash anger and mistrust of Jews confronted both Planck and his good friend Albert Einstein in 1920—Planck considered it a temporary annoyance, but it shook Einstein even as he rose to newfound international celebrity.

Planck was much more concerned with keeping Germany's once-mighty scientific enterprise afloat. He worked without rest, walking two hours to his campus office during Berlin's frequent transit strikes.[10] The professors had returned from war, and the surviving students could again trickle back to their classes, but scarce funding threatened once-mighty German science. The great inflation had already picked up momentum by early 1920. Prices would eventually change by the hour, ever higher. During one of Planck's routine lecture trips in 1923, the cost of lodgings rocketed past his university's travel allotment, so Max opted to spend an evening in the train station's waiting area.[11]

Planck sought a way to keep the expensive enterprise of science afloat without relying on the wobbly Weimar government, punch drunk as it was from economic blows. He joined forces with the chemist Fritz Haber and Adolf von Harnack (ever the pro-science theologian), to found the *Notgemeinschaft*, a private society that could fundraise around the world and then dispense research grants to the scientists at home. The project was a qualified success, ascending to a 1928 budget of eight million marks while winning contributions from General Electric and the Rockefeller Foundation in America, as well as major donations from the Japanese industrialist Hajime Hoshi.[12]

Planck prioritized grant funds for the more theoretical work and especially work on quantum theory, which was fast approaching its enigmatic adolescence. Johannes Stark, ever the laboratory scientist, complained that the *Notgemeinschaft*'s decisions favored "the mathematically oriented Jewish group."[13] Becoming more and more disenchanted with the direction Planck chose for physics, Stark resigned his professorship in 1922, saying he could live on his Nobel winnings. But he maintained a post at the German standards bureau and, teaming with men like Philipp Lenard, ramped up his political meddling. The two physicists would soon pledge their unwavering support for Adolf Hitler, comparing his genius to that of "Galileo, Kepler, Newton and Faraday."[14]

Einstein, unlike Planck, suffered direct attacks. As his star ascended, it sparked the dry kindling within the most resentful segments of Germany;

joining anti-Semites were conservative politicians reacting to his liberal politics, and conservative scientists wanting no part of his challenging theories. Paul Weyland, a German engineer about 10 years Einstein's junior, organized a group dedicated to "Deutsche Physik," a German physics free of abstract mathematical work and free most importantly of Jewish influence. Weyland (allegedly with funding from American businessman Henry Ford), aimed to put on a series of public lectures under the rubric "Study Group of German Scientists for the Preservation of Pure Science."[15] But the series lasted one night. On August 24, 1920, Weyland welcomed a full auditorium to discuss and critique relativity, Einstein's claim to fame. Against the advice of friends, Einstein decided to attend, thinking it might be worth a laugh. But he was shocked at the vitriol, as he heard himself alternately described as a plagiarist, a seeker of fame, and a purveyor of scientific Dadaism.[16]

Planck made no public statements after the event, but he wrote to Einstein that Weyland had spewed "scarcely believable filth."[17] Several colleagues, including Max von Laue, immediately published a joint-letter in Einstein's defense. The ever-passionate Einstein again ignored their advice and wrote his *own* rebuttal as well. Published in the *Berlin Tageblatt*, he titled his long letter, "My Reply. On the Anti-Relativity Theoretical Co., Ltd." He didn't employ subtlety.

> I am fully aware of the fact that both speakers are unworthy of a reply from my pen; for, I have good reason to believe that there are other motives behind this undertaking than the search for truth. (Were I a German national, whether bearing a swastika or not, rather than a Jew of liberal international bent ...)[18]

The ellipsis is Einstein's, and Germany would complete that statement for him in the years to come. To set his own defense, he lists the notable theorists who cite his work, including Lorentz and Planck first. He states that the only physicist of note to oppose relativity theory is Philipp Lenard, who "has yet to accomplish something in theoretical physics." (A decade earlier, Einstein had privately dismissed Lenard as "full of bile and intrigue," which was yet another of his prescient theoretical models for reality.)[19] Near the end of Einstein's letter, he challenges any skeptics to see him the following month, proposing a debate at Bad Nauheim—a small-town corral for a scientific shootout at high noon.

Many physicists found Einstein's bravado and anger just as embarrassing as Weyland's event itself. As friends chided Einstein, he responded with a slightly embarrassed shrug but no apology. "Everyone must, from time to time, make a sacrifice on the altar of stupidity. ... I did so thoroughly with my article."[20] Meanwhile, Lenard felt himself an innocent bystander. Though sympathetic to Weyland, he hadn't attended the ridiculous event but found himself insulted in Einstein's reply. He called for an apology, but Einstein didn't budge. Tensions rose for the Bad Nauheim showdown.

The confrontation took place on September 23, capping a week-long conference. Planck moderated the public discussion between Lenard and Einstein, ensuring civility and decorum throughout. He reportedly cutoff hecklers and dismissed one questioner who referred to newspaper articles to support himself.[21] In the end, those awaiting fireworks were disappointed. Lenard, in summary, said that relativity theory "offends against the simple common sense of the scientist." Einstein remained calm overall, arguing that common sense had shifted along with the advance of human knowledge, even in the time of Galileo.[22] Spectators reported that, "Planck discharged his office of chairman with great ability, strength, and impartiality." And he closed the meeting with the best his Prussian sense of humor could muster, referencing time dilation. "Since the theory of relativity has unfortunately not been able to extend the time available for this meeting, it must now be adjourned."[23]

Einstein caught Lenard at the coat check after the debate, requesting a less formal, more personal conversation. "I silently and forcefully shake my head," Lenard wrote to his diary. "He again, more urgently. I say, 'No, it cannot be done. . . .' Then hat and umbrella finally arrive; I rush away and leave Einstein standing there."[24]

The fallout from Bad Nauheim included Weyland calling the German Physical Society a "rat's nest of scientific corruption," and Lenard sliding further into bitter anti-Semitism.[25] His correspondence with Planck's close friend Willy Wien shows Lenard's behind-the-scenes political efforts. In a series of letters from 1922 to 1925, he calls for an "alliance of the German blooded" to fight the "Judiazation" of universities, warns Wien of the "harmful effects of Planck," and mocks "the little Planck Einstein cult."[26] In 1924, he and Stark authored a public pledge of support to Adolf Hitler, just months after Hitler's failed coup attempt. They wrote that they envisioned, "Founding a new Germany, with Hitler 'beating the drum,' in which the German spirit is not just tolerated again to a certain extent and released from imprisonment, no, but in which the German spirit ... can then finally thrive again and develop itself further for the vindication of the honor of life on our planet which is now dominated by an inferior spirit."[27]

Rumors claimed Einstein might leave Germany, and indeed he would find ready offers elsewhere. "I feel like a man in a good bed," he told a reporter at the time, "but plagued by bed bugs."[28] According their colleague Paul Ehrenfest, Planck and Einstein talked about the idea directly; Planck asked his friend to stay unless things became plainly worse for him. Einstein reportedly replied that he understood his leaving "would be doubly painful at this time of supposed humiliation (for Germany)." The reeling homeland needed him if it wanted to regain its standing with other nations. He was already the only German scientist invited to major postwar international conferences. In the end, Einstein said he felt "bound by the closest human and scientific ties" to Berlin.[29] He would stay for now.

Planck had invested in Einstein early, not only in supporting his first major works, but also in recruiting him to Germany before World War I and before Einstein's ascension to celebrity. Planck placed the still-young man atop Berlin in 1913, as the crown jewel of German physics. Although Planck saw them both as permanent fixtures, it was not to be.

Einstein had originally sought to avoid university life altogether. After his "miracle year" of 1905, he ran away from academic appointments, trying and failing to secure the equivalent of prep school teaching jobs. He worried that a university job would stifle his flexibility and creativity. He finally agreed to his first academic post in 1909.

By then he was vexed by two distinct scientific problems—that they remain so separate vexes physics to this day. One involved his extensions of Planck's ideas concerning physics on the smallest scale where entities defy any normal or comfortable classification. Is it a particle? Is it a wave? In this Lilliputian realm, Einstein would make great and fundamental strides, while also finding immense frustration. The other problem was of an opposite scale, spanning vast stretches of space, and it involved what he called the greatest idea of his life. Solving this problem would make him absurdly well known around the world while putting him in the crosshairs of radical groups in Germany.

This more expansive problem had bothered him for years, as it followed on the heels of his 1905 relativity breakthrough. In 1907, even as Planck completed a relativity-infused update to Newton's mechanics, Einstein fidgeted with an annoying remnant issue. His new relativity had set light speed to be a universal constant, always measured to be the same, no matter who measured it, and no matter their relative speed. He wanted the laws of physics to be equivalent to all observers as well, but it was not yet so. That same year, automobiles began to rumble about the streets of Europe, and electric rail systems rapidly replaced horse drawn buses. Einstein could picture an electric tram passenger. Relativity theory could describe any speed, whether the tram was parked at a stop or moving quickly across town. But as soon as it sped up or slowed down—any time it *changed* its speed—his theory was no longer valid.

Accelerating systems (called "non-inertial" reference frames) were not covered by Einstein's 1905 relativity, and that work came to be called the "special" (as in limited) theory as a result. This limit had to haunt the man who was every bit as committed as Planck to finding universal truths. Accelerating frames are not unusual—all of human reality dwells there. Any object making a circular path is actually accelerating, because it constantly must be nudged to change its direction. So every human accelerates as the Earth rotates, which in turn accelerates as it orbits the sun, and so on. In the end, Einstein's special theory covered only a small subset of natural motion in the universe. A rider may appear to stand still as the tram is parked (neglecting the spin of the Earth for a moment). But even as he hops from the tram's steps to the cobblestones beneath, he leaves the realm of an inertial frame while he falls.

Einstein wondered about a falling person. What about a person standing on the bottom of a rocket ship that fires its engines, moving ever faster and exerting a pressure on that person's feet? A key thought followed closely. And what about a person standing in a room that looks just like the interior of a spaceship, but sits on Earth, with no engines firing and no motion at all? Well, gravity supplies the pressure on that person's feet, and it probably feels just like the pressure experienced in a zooming rocket ship.

What, he thought, would be the observable difference for the astronaut in the accelerating spaceship and the person who sat on Earth, in the room *shaped* like a spaceship? What type of experiment could detect the difference? Einstein's answer became his "equivalence principle": An observer removed from a gravitational field and accelerating (e.g., with rocket engines), should be absolutely convinced that she could be standing in a gravitational field (but pulled to the floor by gravity). At least in her local confines, physics would not be able to tell her if she was rocketing through space or just sitting on a large, gravity-soaked planet somewhere.

The equivalence principle sounds simple and perhaps even logical, but as with the special theory, the implications are drastic. Einstein first confronted the conceptual fallout as he returned his focus to relativity circa 1911. By 1915, this pursuit would result in what he eventually called, "the most valuable discovery of my life" and a "theory of incomparable beauty."[30]

He started a series of thought experiments: setting himself in space ships or elevators, turning flashlights on and off and watching their light beams fly off through space. "I work very eagerly but without much success. Almost all of the ideas that come to my mind have to be discarded again," he wrote to Jakob Laub in 1911. "The relativistic treatment of gravitation is causing serious difficulties."[31] By 1912, Einstein had immersed himself in gravity and a new, expanded type of relativity theory, allowing himself thoughts of little else.

In the summer of 1913, Planck and his colleague Walther Nernst took a train to Zurich and delivered an unusually fetching job offer to Einstein, then 34 years old. Come to Berlin and take an academic post with no teaching required. Your only duty will be to pursue your own research. Come to Berlin for the salary of 12,000 Deutsche marks per year, the highest of any professor in Berlin.[32]

In requesting this sort of salary from the Prussian Academy of Sciences, Planck and his colleagues admitted "their proposal to accept so young a scholar as a regular member of the Academy is unusual" but they argued that this young man presented an even more unusual opportunity, "an especially valuable gain."[33] So stick-like Planck and rounder Nernst both donned formal attire and made their pitch to Einstein.

He didn't answer Planck and Nernst right away. He asked them to escort their wives on a few hours of sightseeing around Zurich, especially taking the funicular tram for its mountain views. And he promised to have their answer waiting for them on the tram platform when they returned. He would hold a rose, he told them: If it was white, his answer was no, but if it was red, he would join them in Berlin. Hours later, as the funicular descended, there was the young genius with dark curls, holding a red rose.[34] Planck had his man, and Einstein stepped onto the main stage of physics: Berlin.

And in 1913, Einstein knew he walked onto sacred theoretical turf as well. Gravity had been the dominion of Sir Isaac Newton for nearly two and a half centuries, but Einstein felt progress at hand, and there was no turning back. The new "general" theory of relativity grew out of Einstein's original relativity that launched in 1905. While Planck loved the first version, he was never as enthusiastic about the more general version, and he only began to budge in 1919, writing to Einstein that he finally embraced it, but not in his heart of hearts—"only as a theorist."[35]

As Einstein moved to join his friend Planck in Berlin, he was still inwardly running thought experiments like the comparison described earlier. We take one person in an accelerating rocket ship, and the other standing in a room on Earth. But now, according to Einstein's equivalence principle, they should see an identical set of physical laws. Any experiment one of them runs should provide the same result when the other runs it. This quickly causes trouble for our conventional views. If the astronaut shines a laser pointer toward the side of his ship, the accelerated motion gives the light beam the appearance of "bending" ever so slightly toward the floor. That's simply because the ship moves forward ever faster, and the straight line of the light ray is left behind, moment to moment. Although this would be incredibly difficult to measure, because light is so fast, the effect is there nonetheless. The equivalence principle now says the person on Earth should measure an identical effect. He stands in his room, turns on a laser pointer, aiming at the wall, and according to Einstein, he should observe a slight inclination of the beam toward the floor. Literally, the laser beam should be warped (very slightly) by Earth's gravity! So in 1911, Einstein first put forward this idea in his paper "On the Influence of Gravity on the Propagation of Light": Rays of light should be redirected by massive planets and stars. Contrary to Newton, Einstein was predicting that a beam of light, with no mass, could and would be *bent* by strong gravity.

To underline the process of science, I should add an extra pinch of historical spice here. As brilliant as Einstein's prediction of light deflection may have been, it wasn't technically original. Following unpublished predictions from England in the 1700s, the German mathematician and physicist Johann v. Soldner published a predicted light deflection in 1804.[36] (Philipp Lenard, in his fierce attacks against Einstein and "Jewish physics," actually used Soldner's earlier prediction to paint Einstein as a plagiarist.) Soldner's prediction gives the wrong numerical result for the deflection of light, whereas Einstein's general relativity gives precisely the correct answer. But the point is that all great scientists—even Einstein and, as we will see, Planck—do borrow, or at least unknowingly revisit, old notions.

Innovative scientific minds can rearrange and flip puzzle pieces in "aha!" moments, but very rarely do they invent new pieces.

To check for starlight deflection, Einstein needed a solar eclipse. Given the moon's assistance in blotting out the sun's glare, a few stars positioned close to the sun in our sky should be visible, at least briefly. And general relativity predicted that the apparent position of those stars would be shifted from their normal positions by the sun's massive gravity (an effect we now call gravitational lensing). His first chance came and went in 1914, in the Crimea region, scuttled by the outbreak of World War I. To Russian soldiers, a trio of Germans, allegedly sent by Einstein with telescopes and cameras, looked a lot like spies. The critical minutes of total eclipse came and went without measurement, more an omen for Europe than an opportunity for physics. (This interruption would be surprisingly critical to Einstein's eventual fame.[37] In 1914, his theory still needed revision, and the first draft was numerically incorrect—photo evidence from the eclipse would have conflicted with his theory.)

Upon Einstein's 1913 arrival in Berlin, most physicists there had to admit they couldn't understand his latest work. Even if they had come around to his first relativity papers, they couldn't follow his methods now. During the red-rose summer, Planck considered Einstein's goal to take relativity to the realm of gravitation. "As an older friend, I must advise against it," he said. "In the first place, you won't succeed, and even if you do, no one will believe you."[38] One world war later, Planck was wrong on both accounts.

In parallel to Einstein, others had been working on the mathematics of special relativity. One mathematician in particular began to describe a natural geometry for Einstein's 1905 work. Hermann Minkowski, one of Einstein's Zurich professors, had shown that simply multiplying time by the speed of light provided a natural *fourth* dimension to pair with the three normal directions of space: up-down, left-right, front-back. In Minkowski's emerging picture, relating the distortions of time and space between one moving observer and another became a simple matter of geometry, and the fabric of the cosmos was *spacetime*.

Einstein was initially not too happy with Minkowski's meddling. He joked, "Since the mathematicians have grabbed hold of the theory of relativity, I myself no longer understand it."[39] By 1912, however, Einstein embraced a more mathematically intense approach for gravity. He began to borrow cutting edge mathematical tools of the time, sometimes enlisting the experts of these fields. These were exotic techniques to physicists of the early twentieth century, and they made Einstein's new work a tough sell.

Einstein understood by now that a more general theory of relativity would be built of geometry—not triangles and straight lines, but a variety with curved surfaces (called non-Euclidean). As he later explained this breakthrough to his son, "When a blind beetle crawls over the surface of a curved branch, it doesn't notice that the track it has covered is indeed curved. I was lucky enough to notice what the beetle didn't notice."[40] To create a system in which light beams naturally curved near massive objects, Einstein recognized the need to give light a good excuse: *curved spacetime*. Brian Greene has described the resulting theory as a "choreography" for the universe's leading stage actors: mass, time, and space.[41] Einstein's choreography dictates that an object's mass distorts the fabric of space-time around it, and the fabric's curvature then instructs the movements of other nearby objects. In the general theory of relativity, Newton's falling apple is not pulled by the Earth—it just follows the steep slope of spacetime near the Earth's surface.[42]

By November 1915, in a flurry of activity, Einstein had a finished theory and a real hook for his skeptical audience. Far from just a beautiful set of mathematics, his new theory resolved a minor but long-standing mystery: the planet Mercury's quirky orbit. Mercury, in its close proximity to the sun, doesn't behave as Isaac Newton's version of gravity would predict, at least in some finer details. But in 1915, Einstein's new theory explained Mercury's orbit perfectly.

"I was beside myself with joyous excitement," Einstein wrote to Paul Ehrenfest.[43] He had satisfied his original misgiving completely. Physics could now explain precisely the observations and experiences of *any* observer, no matter her state of motion. She could sit in a lurching electric tram, or on a spinning planet in a moving cosmos, and rest assured that physics could explain her observations. The more dramatic and public confirmation of Einstein's genius was still to come, as he awaited the next total solar eclipse.

In the war years, both Planck and Einstein worked through their bleak circumstances, illnesses, and tragedies, to somehow make critical contributions that are still of use today, if somewhat overlooked in their legacies. And in this period, they actually revisited one another's work from years earlier. It is as if, as they played music together in those times, they wondered what it would be like if Max took up Einstein's violin and Einstein tried his hands at the piano. Among other projects, Planck returned to Einstein's second "miracle year" paper of 1905 and the problem of Brownian motion, the random herky-jerking of a microscopic particle. In 1917, Planck enhanced Einstein's statistical method for predicting the particle's future locations. This work, combined with that of Lorentz's student Adriaan Fokker, is today described by the Fokker-Planck equation in a flourishing field called stochastic dynamics. Meanwhile, because turnabout was fair play, Einstein delved into Planck's own seminal 1900 discovery, centered on black-body radiation. In his 1917 approach to thermal radiation, Einstein separated an object's automatic thermal emission of light from its emissions that only arise after external prodding. The study of this so-called *stimulated* emission led directly, over time, to the invention of the laser.

The next opportunity to test Einstein's general relativity arrived after the war. In 1919, the British astrophysicist Arthur Eddington led an expedition to Principe Island off the West African coast. The day started with heavy rain, but for the critical two minutes of total eclipse, the clouds parted, shutters snapped, and starlight deflected exactly as Einstein's 1915 calculations predicted. Eddington telegraphed his results home, and news quickly made its way to Planck and Einstein, via their Danish friend Hendrik Lorentz (Figure 9.1).

Planck sent his congratulations.

> Dear Colleague, I don't know when I am going to have the opportunity to talk to you at leisure, but I do know that I cannot postpone telling you until then how deeply and how heartily pleased I was about the news contained in Lorentz's telegram. Thus the intimate union between the beautiful, the true, and the real has once again proved operative. You have already said many times that you personally never doubted the result; but it is beneficial, nonetheless, if now this fact is indubitably established for others as well . . . In the meantime, cordial greetings from your devoted servant. M. Planck.[44]

It is worthwhile to note how little we understood of the larger universe in 1919. Much was still to come. Just as the war impacted general relativity's vindication, it also paused the career and discoveries of a precocious American astronomy student named Edwin Hubble, who stopped his PhD work to enlist in 1917. So at the time of Einstein's triumph, and even as Einstein began to apply general relativity to the very shape and structure of the universe itself, young Hubble was returning from France, having not yet discovered the existence of other galaxies. And he had not yet made

Figure 9.1. Albert Einstein, Hendrik Lorentz, and Arthur Eddington circa 1919. Courtesy AIP Emilio Segre Visual Archives.

the existentially terrifying discovery that the galaxies were flying apart from one another, just like so many fragments from a great explosion, a "Big Bang."

The dramatic eclipse experiment captured a war-weary public's imagination and garnered front-page international headlines. Articles assured readers that the theory itself was beyond the grasp of most humans, but no matter: Einstein would soon be a household name, a vision in film reels, and a new kind of quotation engine for the press.

Why did Einstein become one of the most famous people of the twentieth century? Most will note his natural enjoyment of the spotlight, his good and distinctive looks, and his view that scientists must engage the public. One can also point, as Thomas Levenson does, to timing. Technology had just evolved to providing fast and global news, radio interviews, photographs, and film clips. Given the silent newsreels of the day, Einstein's ready smile and penumbra of hair helped sell more "a persona than a person."[45]

And as biographer and friend Abraham Pais wrote, part of the fascination with Einstein was, for most people, a pure type of respect and awe, unsullied by any understanding for his work itself.

Colleagues in Berlin were aghast at so much attention for a scientist. They clicked their tongues when Einstein allowed his photo to appear in a book on relativity. And when a tabloid writer convinced Einstein to let him write a version for the masses, including several long interviews, some of his closest friends intervened. In writing an appeal for moderation, the physicist Max Born said the issue of decorum "concerns everything dear to me (and Planck and Laue, etc.) . . . in these matters, you are a little child." And Born's wife wrote to Einstein that his behavior smacked of (the dreaded) "vanity."[46]

In 1921, Einstein began touring the world with his cousin and second wife Elsa, including a ticker-tape parade in New York City. While the world adored a Jewish German scientist, one who seemed to enjoy or even goad the attention, the same world simultaneously showed its disdain for vanquished Germany. Against this backdrop, mixed with growing anti-Semitism and disgust with the Versailles treaty, Weyland and Lenard launched their attacks in 1920.

Max Planck watched his friend's fame ascend and tried to protect his scientific reputation at home (as in the Bad Nauheim meeting of 1920). Their personal bond had grown during World War I, despite political differences. Perhaps the musical collaborations and Einstein's open sympathy for the Planck family sorrows tugged at a reserved Prussian heart. "Planck loves you," Elsa Einstein wrote to her husband in 1921.[47] To bolster Einstein's image in the *scientific* community, to pull him from parades and back into the temple of science, Planck invited Einstein to give a keynote address at a major scientific meeting: the 1922 centennial celebration of the Society of German Scientists and Physicians. Einstein happily accepted the invitation.[48]

Planck helped heal his own deep wounds by playing mother hen in these years. In addition to Einstein and two young, motherless granddaughters, Planck worried over his most dear friend, his surviving son Erwin. A still-painful war wound no doubt resonated with his open emotional scars, and like Grete and Karl before him, he now struggled with depression. Erwin had always shared the strongest bond with Emma, and the silence of her once lively voice must have pained Erwin incredibly. Gone were her beautiful letters and ever-positive tone. When the siblings had learned of Karl's death by shrapnel, they exchanged several letters. "Now we can only show our love by letting him live and act within us," Emma wrote to captive Erwin. "We must hold on together, we three."[49] But just a few years later, the three were reduced to one. In 1922, Erwin moved back to Wangenheimstrasse, presumably with strong urging from Max, to recoup in the family nest and spend more time with his father. To better battle his angst, Erwin began reading a new German novel, *Siddhartha*. Hermann Hesse's book traces the fictional life of a man in ancient Egypt. He renounces his material comforts and seeks out the Buddha in person.

On the morning of June 24, 1922, if Erwin and Max enjoyed an early coffee on the veranda, they might have spilled their cups at a sudden flare of machine-gun fire. Germany's foreign minister Walther Rathenau lay dead nearby. Just blocks from Wangenheimstrasse, a radical group of students ambushed Rathenau as he started to work in his convertible. Like Einstein, Rathenau was a German Jew, but unlike Einstein, he was strongly nationalistic, and he believed the way to crush anti-Semitism was to have German Jews assimilate. Rumors soon spread of a political hit list with the group's next targets, and it supposedly featured Einstein's name.

The Weimar government ordered a national day of mourning with all businesses and classes suspended, but Philipp Lenard and his ilk refused to mourn.[50] He went on with his lectures as millions gathered in the streets of Berlin. At the same time, Lenard's ally Johannes Stark drafted *The Current Crisis in German Physics*, a manuscript detailing the pervasive and caustic effects of Jewish and mathematical physics. He named Einstein as a primary villain and advised physics students to head for work outside of the universities, "like Johannes Stark" had done.[51] When his book appeared the following year, Max von Laue soon published a review. "All in all, we would have wished that this book had remained unwritten, that is, in the interest of science in general, of German science in particular, and not least of all in the interest of the author himself."[52]

Meanwhile, Einstein didn't need Stark's ravings to understand the changing mood. The danger felt especially real, because he had known Rathenau personally. Einstein cancelled his upcoming relativity talk and fled to Kiel, sending Planck his regrets.

> A number of people who deserve to be taken seriously have independently warned me not to stay in Berlin for the time being and, especially, to avoid

> all public appearances in Germany. I am said to be among those whom the nationalists have marked for assassination. Of course, I have no proof, but in the prevailing situation it seems quite plausible. The trouble is that the newspapers have mentioned my name too often, thus mobilizing the rabble against me. I have no alternative but to be patient—and to leave the city. I do urge you to get as little upset over the incident as I myself.[53]

"We are lucky to have come so far," Planck replied to Einstein, "that a band of murderers . . . dictate the scientific program of a purely scientific body."[54] He was incensed by the threats to the most important of German physicists. Over the next two years, he tried to discern the exact nature and origin of the threats against Einstein, communicating with the police and even corresponding with officials working in insane asylums. But he learned nothing.[55]

Einstein decided to take further time away from Berlin. He and his wife embarked on six months of travel (including his first visit to the Holy Lands, and a series of talks in Japan). In November 1922, with Einstein half a world away, Germany received word that he would receive the long-delayed 1921 Nobel Prize in Physics. Stark, no doubt feeling his own prize downgraded in the bargain, typed an indignant letter to the Nobel committee, chastising them for "perpetrating such a fraud."[56]

Planck gave a talk in the summer of 1922 decrying the pessimism that gripped his beloved nation, even as inflation binged on German currency. And while his own home was stalked by ghosts and one forlorn son, Max looked to celebrate good news where he could find it. One of the only bits came via Lise Meitner, who that autumn became the first woman lecturer at the University of Berlin. Planck had helped engineer her history-defying rise in rank, and she wrote a glowing letter thanking him for his efforts and the new honor.[57]

By the end of 1923, Max Planck could celebrate two additional pieces of good news. Einstein would return to Berlin after collecting his Nobel Prize. Meanwhile, Erwin, now 30, had found his true love, a lively and sharp 20-year-old named Nelly Schoeller. They married in December, and General von Schleicher stood as their honorary witness. Celebrating the newlyweds, the Planck house enjoyed an optimistic holiday season for the first time in many years.[58]

If the Plancks wanted to avoid darker conversation over their Christmas dinner, they probably steered clear of current political events. But Erwin now worked within the Weimar government, and he would have carried a cloud of fresh worries with him. Adolf Hitler and 600 members of the NSDAP had just attempted a coup in Munich. As Max wrote to Einstein at the time, it was "as criminal as it was short-sighted."[59] If someone mentioned the "beer hall putsch" that Christmas, perhaps Max tried to reassure Erwin and Nelly. Hitler, in custody and awaiting trial for high treason, was then a man as defanged as he was deranged.

10

July 1944—A Celebration

As Planck recovered from surgery, the Allies took Rome on June 5 and swarmed the beaches of Normandy on the sixth. Most Germans could see an end to the war methodically approaching, despite Nazi propaganda.

In early July, Max Planck donned his once-familiar tuxedo and traveled to Berlin for a scientific celebration. The Prussian Academy of Sciences would, for one night, try to turn away from the darkness surrounding them and celebrate an anniversary. The Reich's top physicist and director of their atomic research program, Werner Heisenberg, organized the event. Heisenberg collected Max Planck at his Berlin hotel and attempted to drive them to the banquet hall, but nothing in the wrecked landscape was familiar. Neither of them could recognize the streets of Berlin. After asking for help, they finally found the correct address, but once more thought they were lost. "We ended up with our car in front of a giant pile of rubble with bent iron rods and concrete blocks," Heisenberg recalled. After further inquiries, they were shown a path through the dusty, jagged debris, navigating the ruins of the empire, climbing and squeezing their way to an open door. Inside, the surviving banquet room was lively and intact like a sharp memory. When Planck entered, the assembled grew silent. "Each greeted Planck with worship, and you could see so clearly how much love streamed to this man," Heisenberg said. "And you could also feel that he himself was happy to see the familiar faces once again. The string quartet began to play, and for an hour or two the old times returned, the cultured Berlin where Planck, of course, was the leading figure." A large gathering of friends and remnant scientists welcomed the guest of honor and toasted the fiftieth anniversary of Geheimrat Professor Doktor Planck's induction to the Prussian Academy of Sciences. At the lead table, Max was particularly pleased to find himself seated next to Erwin.[1]

Why, 50 years later, had Planck become such a pervasive scientific name? How had the young student showing great aptitude for everything, but superlative ability in nothing, risen to such influence?

Though he possessed a remarkable mind, it is difficult to label Planck a genius in the end. "Unfortunately," he later said, "I have not been given the capacity to react quickly to intellectual stimulation." Planck had never been at the top of his class as a student—rather, his teachers noted his winning people skills.[2] What Planck did possess as a scientist was an uncommon brand of focus, an unmatched mental rigor. He continually honed his concentration's sharp edge in the furnace of faith. He knew in his bones that nature operated by some logical system, and he devoted himself to uncovering that system, no matter the obstacles or consequences.

The transformation from a solid but not-necessarily-distinguished physicist to the enshrined Max Planck began sharply in 1894. His former professor and friend Hermann von Helmholtz nominated Planck for full membership to the Academy. He claimed that Max, then 36, had made substantial contributions to physics. Most notably, Helmholtz wrote, Planck had made his mark on the field of physical chemistry *without* the chemists' deplorable dependence on the concept of atoms.[3] Planck had used pure thermodynamics, assuming only continuous matter (no grains or atoms), to probe with his pen and paper everything from phase transitions to the behavior of fluids conducting electricity. His entire career to 1894 had been devoted to thermodynamics, with special allegiance to the second law and entropy.

In truth, Max Planck's thinking was changed by working on and thinking about problems of chemistry. Quietly and methodically, he had come to accept "atomism," unlike most other physicists. The switch to embracing atoms was the first of many major shifts in his thinking, and to his great credit, this flexibility was a hallmark of Planck. So many academics sink roots into their favorite topic and convictions, but Planck was able to move to a more sensible spot, following the sunshine and the most fertile soil.

In 1882, the young physicist held a common view. Atomic theory, he said, "will ultimately have to be abandoned in favor of the assumption of continuous matter." In the same year, he ended his paper "Vaporization, Melting and Sublimation," by noting, "The mechanical theory of heat is incompatible with the assumption of finite atoms."[4] Statements of this kind were music to the establishment's ears. The idea of "continuous matter" is difficult to take seriously in the twenty-first century, but it passed the sniff test for many previous centuries: Most scientists assumed that given a sharp enough and small enough knife, they could always cut a tiny object into ever smaller pieces. A huge advantage of this assumption was mathematical, as Newton's calculus could reign supreme in describing continuous matter. But in 1884, the Swedish chemist Svante Arrhenius made a breakthrough proposition to describe the electrical properties of certain solutions. He suggested that atoms drifting in water could become ionized, and thereby electrically charged. In the same year, Jacobus van't Hoff published *Studies in Chemical Dynamics*, similarly grounded in the assumption of atoms. And so by 1887, as Planck toiled within these topics, he had changed his mind. And in 1890, he wrote to a colleague that physicists had no choice: If they wanted to examine certain problems, they had to accept atoms and molecules as a given.[5]

In 1894, Max was already a full professor (rank of *Ordentlicher Professor*). The ascension to the Berlin Academy must have felt like cresting one of the mountains he confronted every summer, and at last he found himself embraced by the establishment. His reputation was solid if not spectacular. In five years, he had steadied himself and his family in Berlin. He and Marie celebrated the first birthday of their last baby Erwin. The city itself was abuzz with new electric lighting and a growing number of telephones. Amazed audiences watched Ottomar Anschütz project silent moving pictures of athletes and soldiers trotting their horses. Berlin's well-to-do tinkered with new technologies and anticipated more. Students began filling the seats of science courses like never before.

At the same time, German physics suffered a sudden leadership vacuum. By the end of 1894, both the renowned Hermann von Helmholtz and also the rising genius Heinrich Hertz would be dead.

As described earlier, Max Planck was poised for a new challenge. He was now versed in the problem of black-body radiation, which hinted at universal truths, and in looking around Berlin he saw the world's leading experts collecting black-body measurements, just waiting for a theory to explain their results. Most importantly, he had identified what he considered the single most important problem in physics: reconciling thermodynamics with mechanics. In the former, his home turf, one could define an arrow of time. As entropy gradually increases, moments tumble irreversibly on the heels of prior moments, all according to the second law. But the new "gas theory," a microscopic version of mechanics, didn't fit with the second law. This *statistical* approach of Maxwell and Boltzmann gambled with reality—it placed odds on different configurations of jittering, colliding, and spinning gas molecules. Here, physicists could discuss an *average* quantity for trillions of tiny particles, or a spread of possible options. Instead of one exact snapshot of a system sliding irreversibly and necessarily to the next, a system in gas theory rattled and hummed from point A to point B, most probably, but there was always a chance it could find point C or D instead. And sometimes, moving from point A to point C or D could rupture the second law. In 1895, Boltzmann wrote that, "The probability of such cases is not mathematically zero, only extremely small." Max, on the other hand, felt great loyalty to the ideas of Clausius. "At that time, I thought the principle of entropy increase was valid without exception, like the principle of energy conservation," he wrote late in life, "while for Boltzmann this principle was only a law of probability and thus subject to exceptions."

Max had a difficult professional relationship with the more emotional Boltzmann. In the early 1890s, despite his conversion to atomism, he still thought that much of Boltzmann's work was, "inadequately rewarded by the fruitlessness of the results gained."[6] In essence, he said gas theory looks very impressive, but it's all sound and fury so far. The sensitive and mercurial Boltzmann took offense, and to Max's astonishment, he publicly questioned Planck's judgment during a subsequent conference.[7]

Prickly interactions aside, Planck was now hooked on the problem. How could he reconcile the beloved second law of thermodynamics with the sensible and increasingly successful statistical approaches? Even while the number of people following the dilemma could have been tallied on one hand, Max relished its relative obscurity. "It was an odd jest of fate," he wrote later. "The lack of interest of my colleagues ... turned out to be an outright boon.... As the significance of the concept of entropy had not yet come to be fully appreciated, nobody paid any attention to the method adopted by me, and I could work out my calculations completely at my leisure, with absolute thoroughness, without fear of interference or competition."[8]

Planck turned his full focus to the mysterious and universal curves of black-body radiation, which he intuited as the perfect battleground for thermodynamics and the new mechanics. Thermodynamics would be key since the brightness and colors of the emitted light depended precisely on the temperature. He would also need some of the new techniques of Boltzmann and Maxwell. The molecules of the black body roiled in some sort of statistical reality, with various energies and vibrations, and somehow they gave rise to the exact spectrum of emitted light. By analogy, imagine a crowded stadium at a football game. If we wish to completely understand the roar emanating from the stadium, a physicist can start by building a model of the fans in the seats. She gives the fans a statistical distribution of standing or sitting, screaming or clapping, and different volumes throughout. The physicist could start to approximate the sound that we then hear from some distance. In this sense, the thermal radiation spectra were taunting Planck. Why was the spectrum (the roar) independent of the black-body's material (the exact stadium, location, and crowd size)? And why did the spectral curve of light from these bodies have exactly one shape and no other? His attraction for the universal pulled him to thermal radiation.

As he entered the problem, Planck would for the first time need to confront light in detail. If there were two things Planck felt he could count on in 1894, they were the second law of thermodynamics and the newly confirmed theory for the propagation of light. The esteemed Heinrich Hertz (who had been offered both of Planck's jobs before Planck), had published his evidence for electromagnetic waves in 1888, and Planck had started following his work closely.[9] "Dear colleague, Sorry to take your time but it's a quick question," Planck wrote to Hertz in the summer of 1890. "I read your article 'Electrodynamics for stationary bodies' with great delight." Planck went on to ask if perhaps Hertz had made an error with a minus sign, and hadn't he made a typo in equation 8? "I really hope I didn't take too much of your time." Planck was notorious for this kind of exacting reading, and it helped make him an excellent editor of *Annalen der Physik*.

He approached the black-body problem in his usual methodical fashion. He began by picturing a hollowed out pocket inside a material—a cavity—just like his old professor Kirchhoff. In 1860, Kirchhoff published his original examination of thermal radiation. He possessed an incredible ability to assimilate vast swaths of information and render an insightful, unifying summary. And he had done so again with the spectral radiation emitted from objects, noting from his own experiments that the ratio of absorbed and emitted light from an object did *not* depend on the material itself; he coined the term "black body" (as well as "gray body" for the not-quite-perfect black body), and declared that a "universal function" underwrote the thermal radiation.[10] (It is still astounding that, at such an early time, Kirchhoff postulated a universal character of thermal radiation, despite inadequate and incomplete experimental evidence to support the notion.)[11]

Planck in his calculations assumed the cavity was a perfect sphere and had *no* air in it—the pocket held a perfect vacuum, with no debris or even a single molecule of gas. The entire problem then dealt with the electromagnetic waves (light), inside the cavity and their interaction with the cavity walls. Since the actual material of the walls would not matter—again, mud, chocolate, and metal would behave the exact same way for thermal radiation—Planck used the most generic version of matter. He assumed a series of "resonators" on the inner surface. Though we might be tempted, with modern eyes, to read specific physical meaning into Planck's resonators, he most probably just sought the simplest possible building block for his theoretical black body.[12] He borrowed the notion of an electric resonator from the work of Heinrich Hertz, citing one of Hertz's papers from some six years prior.[13] So while the Hertz resonators had an electrical character, Planck might have shrugged off any question about their real nature. Interestingly, the idea of oscillating molecules giving off radiation had been proposed years earlier (in English, by James Clerk Maxwell), but Planck had probably not seen it.[14] Eventually, Planck and others would speculate that the resonators might be electrons, but at the beginning of his work, he wrote of no specific physical meaning for them.

Planck's resonators each had a characteristic frequency, meaning that each one could interact with one and only one color of light. If we go back to the analogy of a football stadium, Planck was assuming each fan could only absorb and emit one tone. As with many initial models in physics, this assumption is about as realistic for the walls of a cavity as it is for football fans. The aim of Planck's approach was to model the way these abstract electric resonators took in and then sent out light into the empty pocket. (His first lecture on the topic, to the Prussian Academy of Sciences, was entitled "Absorption and Emission.") He hoped that the eventual radiant spectrum he computed would simultaneously match observed laboratory measurements *and* prove that the spectrum maximized the cavity's entropy, confirming and upholding the second law of thermodynamics.

Through the first several years, he tried to conquer the problem using classical and continuous techniques. His first black-body papers built a framework with few complications, but then slowly increased the model's complexity, from one paper to the next. He aimed to show that no matter what kind of light might fill the cavity initially, that the activity of the resonators would eventually yield the observed black-body radiation. He wanted to uncover a one-way arrow of physics from *any* starting point to the final one, so that he could explain how any object would arrive at the same experimental output. But doubts began to encircle him in late 1897.[15]

"Any unidirectionality which Hr. Planck finds," Boltzmann wrote in a devastating critique, "must . . . derive from his choice of unidirectional initial conditions."[16] Boltzmann's point was that Planck's attempts to show a universal and unstoppable series of events were not universal at all. Boltzmann correctly detected that Planck's theory had hidden training wheels. It wasn't quite ready to balance on its own. Nature clearly could start anywhere and arrive at the same endpoint. But Planck's theory needed a carefully selected starting point or it tumbled over. A robust theory should not be sensitive to the so-called initial conditions, and every starting point (A, B, or C) should lead to point Z, the black-body spectrum seen in laboratories. After initially disputing Boltzmann's zinger, Planck conceded the point by the end of the year. He had to try something new.

Planck knew and respected Boltzmann's work even though he struggled to assess its utility. While Boltzmann embraced probability calculus as a favorite tool, it would have struck most physicists as esoteric, extreme, or even confounding in the 1890s. But by 1898, Planck was warming to Boltzmann's ideas.[17]

That year, Planck adopted a more statistical approach to the thermal radiation problem, computing *average* quantities over the entire spectrum of possible resonator frequencies (and the resulting light colors). He introduced a notion of "natural radiation," which was the electromagnetic version of Boltzmann's "molecular disorder" from his first gas theory papers. Planck wrote a true capitulation in 1899, published in a summary article, where he bluntly admitted that Boltzmann and gas theory offered the best path for understanding his sacred entropy and the second law of thermodynamics.[18]

Meanwhile, something was amiss in Berlin physics laboratories. For a few years, the empirical side of black-body radiation had known relative peace. In 1896, Willy Wien (Planck's friend and eventual co-editor of *Annalen der Physik*), had concocted an equation that made a good fit to all black-body data from the labs. Most physicists accepted Wien's equation as an empirical approximation to the truth, if not the gospel truth. But in 1899, new measurements at longer wavelengths (e.g., looking at infrared light) drifted up and away from Wien's predictions. Infrared light was shining with unexpected intensity from the hot cavities, so Wien's equation was clearly not the one used by nature. New experiments pushed further into the infrared, and they arrived on Planck's Berlin desk every few months in 1899 and 1900. Planck began thinking about a new empirical fit to replace his friend's. Frustrated by years of trying to build a new theory from scratch, he decided to work backward. If he could intuit the correct equation for nature's thermal radiation, that would provide a precise compass heading as he navigated the underlying physics.

Turbulence accompanied the weeks leading to his breakthrough. The fall of 1900 witnessed the departure of two German Wilhelms. In August, after labor leader Wilhelm Liebknecht died, the unseen working classes emerged as more than 100,000 choked the streets of Berlin with a solemn funeral march. Even as the Emperor denounced Liebknecht as an enemy of Germany, the march of so-called "iron silence" chilled the city and underscored the vast class divisions of the time. In September, a much closer Wilhelm passed next, as Max's father died in Munich.

Having paid his respects in Munich, Planck returned to his packed lectures in Berlin. In the weeks following his father's death, Planck had a flash of insight. Was this back in his Berlin office, during one of his lectures, or even on the train returning from Munich? Whatever the setting, Planck now clearly beheld the mathematical shape nature had been yearning to reveal. The elegant form struck him, and he wrote out a new equation that seemed to avoid the problems of Wien's earlier guess. On October 19, Max briefly presented his results at a regular meeting of the German Physical Society. He took the floor not with a scheduled talk but with a last minute "discussion remark."[19] His spoken comments consume barely more than two pages. He admits, "I have finally started to construct completely arbitrary expressions for the entropy" of the resonators. When he then writes his result for the audience, he says, "As far as I can see at the moment, [this equation] fits the observational data, published up to now, as satisfactorily as the best equations put forward for the spectrum."[20]

$$E = \frac{C\lambda^{-5}}{e^{c/\lambda T} - 1}$$

This gives exactly the energy E (and hence the intensity) of radiation one would measure at the wavelength λ for a black body at a temperature T. The big C and little c are just constants that help fine-tune the equation to match the measured data. Such constants are the measurements a physics tailor makes when fitting a body of data with an outfit of mathematics. Planck knew he would have to carry out a real fitting for the values of *C* and *c* at some point.

How did his audience react to the unveiling? Presumably, the few interested parties (like the experimenters Pringsheim, Kurlbaum, and Rubens), scribbled this new formula in their notebooks. The following morning, an excited Rubens showed up at Planck's door holding some pages from his lab. All of his measurements, including the latest in the infrared, fit Planck's new curve like beads on a string. Rubens was convinced Planck had found it, nature's thermal voice.

Other colleagues were less convinced. Wien, who heard about the talk from afar, wrote a letter saying he didn't want to be offensive but that the new equation was surely a significant blunder. Planck replied that he was sure they could resolve their differences and that Wien's merits in the field would probably be elevated further in time, not diminished. He suggested they meet over the Christmas holidays.[21]

Meanwhile, Planck was now left with the million-deutsche-mark question: Where did this nice equation come from? His new formula, "had merely the standing of a law disclosed by a lucky intuition," he wrote later. "On the very day when I formulated this law, I began to devote myself to the task of investing it with a true physical meaning."[22] He set about finding a way to justify the new equation by deriving it from fundamental physical principles. He had to turn his guess into a genuine theoretical idea.

"After a few weeks of the most strenuous work of my life, the darkness lifted and an unexpected vista began to appear."[23] Planck saw that to get his derivation on the right path, he needed to root around in Ludwig Boltzmann's statistical toolbox. Planck pulled the crucial mathematical pieces from a relatively old paper of Boltzmann's, 1877's "On the Relation between the Second Law of Thermodynamics and the Theory of Probability."[24] Boltzmann had enumerated the possible configurations for a bunch of gas molecules using a heavy dose of combinatorics, a type of mathematics examining all the various ways that an array of objects (gas molecules, socks, volleyball players, poker cards, etc.), might be combined. For example, if we have a huge jar of marbles, half red and half blue, and we randomly grab three marbles from it, what are the odds that all three are blue? The answer is approximately one-eighth, or a 12.5% chance. The underlying combinatorics required us to consider all the possible three-marble configurations, such as red-blue-red, blue-blue-red, and so on. There are eight possible combinations and only one gives all blue marbles. Of critical relevance to Planck's story, combinatorics is a branch of *discrete* mathematics, with single and separable objects like molecules or marbles. This chunky type of math had no business associating with the continuous light waves arising from a hot cavity, did it? But Max was desperate, and only combinatorics led to his new formula.

In December 1900, he presented his new proof and his new theory. Historically speaking, physicists consider this the proper birth of quantum theory. His audience was reportedly unimpressed, and they struggled to follow Planck through his new forest of mathematics. To obtain his October equation, Max divided allowed resonator energies into parcels. "We consider, however—this is the most essential point of the whole calculation—*E* [energy] to be composed of a well-defined number of equal parts."[25] Once he could treat the resonator energies like a set of socks, cards, or molecules, he could use Boltzmann's methods from 1877, and then march in a sensible way to the new black-body equation.

But the new "essential step" was a radical and subversive one, the first time that anyone had considered treating sacrosanct *energy* in a gritty way. Just as some more progressive physicists had talked about a chunkiness for matter (atoms) and electricity (charges), Planck had quietly submitted a similar granularity for energy. In 1900, he didn't speak of it as a new type of physics—he didn't even use the word "quantum." He said later that this historic step was "a purely formal assumption and I really did not give it much thought except that no matter what the cost, I must bring about a positive result."[26]

To imagine Planck's leap here, consider measuring the heights of children in a large elementary school. We typically measure them on a rough continuum, where children can range easily between, say, three feet and five feet tall. Planck was saying, well, I can describe the heights shown universally by elementary schools all over the world if I just assume children can be exactly three feet high, four feet high, and five feet high, with nothing in between. He didn't seem to believe this to be the underlying truth, in 1900, but it worked mathematically.

Planck thought he had successfully preserved classical physics and solved the riddle of black-body radiation. He was excited, to be sure, and he could spy success, but there is nothing to suggest he sniffed the fire he had just started. Within 30 years, quantum physics would completely rewrite our view of light and matter, largely from the pens of people like Albert Einstein, Niels Bohr, Werner Heisenberg, and others. Through most of those years, Planck would play the role of a worried parent asking everyone to slow down, to be careful please.

Planck's derivation itself, we know now, is fundamentally flawed—not in the assumption of granulated energies, but in his argument for their existence. There is no path from classical physics to the exact black-body spectrum. Younger scientists like Einstein and Paul Ehrenfest would be among the first to poke, prod, and criticize Planck's black-body mathematics, and when Planck eventually returned to the problem, he took a new approach to the derivation. It came to be known as the "second theory" with papers spanning the years 1911–1913. Here, he still hoped to avoid letting the new quantum idea leak into beautiful and continuous light waves. (Interested readers should refer to the appendix, where I give a more modern and proper derivation of the thermal energies radiating from any object, just based on its temperature, and I touch on Planck's second theory, which launched the concept of "zero-point energy" in the universe.)

The year 1900 would become a retrospective pinnacle for Planck and the new physics. Later, Erwin recalled a late autumn father-and-son garden walk. He said his father quietly announced a breakthrough to his seven-year-old son. Max claimed that his recent work could go down among the greatest discoveries in physics.[27] Erwin's story doesn't mesh with what we know of Planck's personality and humility; Erwin could have been shuffling distorted memories of his distant boyhood. But perhaps Max allowed such a grandiose statement for his son's ears only. Even there, the existing letters show no tendency to boast. There was only one part of his work that could have created such confidence and excitement. In deriving his new formula, he illuminated two fundamental constants, k and h. The former came to be called Boltzmann's constant, and the latter was known almost immediately as Planck's constant. While he downplayed the details of his curve's mathematical *derivation*, the resulting *constants* struck Planck as the real breakthrough. He spent the last quarter of his December lecture on their implications.

The unveiling of these fundamental constants, derived purely from Max Planck's mind combined with the available laboratory data, has only become more impressive with time. The values he unearthed for k and h are within just a few percent of their modern confirmed values. More importantly, he immediately mapped where they would help solve other scientific problems. Planck showed that k was the gateway to calculating exact entropies from a Boltzmann-like *statistical* framework. In practical and immediate application, Planck deployed *k* as a type of numeric glue to strengthen the bond between thermodynamics and the field of chemistry. He used *k* to improve the contemporary values of Loschmidt's number (the number of molecules in a cubic centimeter of gas), Avagadro's number (the number of molecules in one "mole" of gas), and the elementary unit of electric charge (the charge on the electron). All of this appears in the last half page of his December 1900 paper. "If the theory is at all correct," he wrote, "all these relations should be not approximately, but absolutely, valid."[28]

Planck's stance on the constant *k* would evolve over the years. After Boltzmann's suicide in 1906, Planck led a group of physicists in calling it "Boltzmann's constant." The historian John Heilbron suggests Planck simply made paid homage, an intellectual debt, to Boltzmann.[29] However, as the years passed he seemed to regret being short-changed of credit for the monumental breakthrough that *k* represented. He sometimes wrote of it as "the so-called Boltzmann's constant," and he later published a somewhat petulant line. "Boltzmann never introduced this constant, nor, to the best of my knowledge, did he ever think of investigating its numerical value."[30]

And though he made fewer calculations with his own constant *h* in 1900, it has proved every bit as important to physics. As Einstein wrote shortly after Planck's death, without *h*, "it would not have been possible to establish a workable theory of atoms and molecules and the energetic processes which govern their transformations."[31]

Planck summarized his derivation in a 1901 *Annalen der Physik* paper, but then turned away from black-body radiation for nearly a decade. He considered the black-body issue resolved, aside from watching the new lab results continue to fit his curve. Otherwise, he turned most of his attention to studying the dispersion of light rays from an object's surface—this was more a study of reflection than black-body emission. Here, he could tinker with new ideas from Hendrik Lorentz in which tiny, identical "electrons" were said to permeate solid matter.

And if he did expect fame from his black-body success, as he allegedly said to Erwin, he must have been frustrated in the years that followed. From 1901 to 1904, only a handful of other authors referenced his work, noting that his equation fit the available data but saying little else.[32] In the early twentieth century, physics was full of more inviting topics than black-body radiation, and besides, Planck had relied on the unpopular ideas of the unpopular Boltzmann.[33] The first substantial action on Planck's breakthrough came in 1905 from the mind of Einstein, but Planck considered that work a bit of wrong-headed excess from an overly exuberant kid. Planck and his colleagues largely ignored the young interloper's suggestion of "light quanta" (what we now call photons).

It was that year that Planck and Marie moved their family to the suburb of Grunewald, just as he began to review Einstein's string of five seminal papers. Einstein's very first *annus mirabilis* paper concerned a strange experimental observation called the "photoelectric effect." Others had noted that shining ultraviolet light onto a piece of metal created a spark—what we would now call an ejection of electrons. This effect powers the photovoltaic materials in solar panels. Classical theories, using light as a wave, could attempt to explain this phenomenon—the waves must surely "shake loose" the electrons, or something like that—but they came up short.

Einstein found a way to explain the effect with just one assumption added to Planck's 1900 paper, but it was radical, even ridiculous to some. He asked, what if light itself *literally* appears in chunks of energy, measured in scoops of Planck's new constant h? If light comes in little bundles, then the photoelectric effect would make perfect sense. As little energy bundles hit a metallic surface, billiard-like collisions between the light quanta and the electrons would ensue. Einstein's model involved just a simple bit of algebra and explained the observed experiments flawlessly.

By proposing such a mechanism for light, Einstein knowingly picked at a painful scab for physicists. The debate about the nature of light had consumed most of the last 200 years, with the wave picture triumphing over the particle picture by the late nineteenth century. Most physicists, especially most established physicists like Planck, had no will to revisit the debate.

In fact Planck's reaction to Einstein's idea of an "energy quantum" for light is intriguing. While he embraced and even championed Einstein's special relativity immediately, he thought the very same genius had made a brash rookie mistake when it came to the photoelectric effect. It was out of character for Planck to ignore evidence this strong and to dismiss an idea so simple and logical. But this was *his* intellectual turf. Special relativity confronted years of frustrated work from the Dutch physicist Hendrik Lorentz and the French polymath Henri Poincaré (who were both very skeptical of Einstein's relativity for the rest of their days). But this idea for the photoelectric effect poked its quantized nose into Planck's business. As flexible as Planck had been in his science, he remained skeptical of light quanta for years to come. In 1913, while writing the otherwise glowing justification of Einstein's Berlin job offer, Planck couldn't help himself: "That he [Einstein] might sometimes have overshot the target in his speculations, as for example in his light quantum hypothesis, should not be counted against him too much."[34] Thirteen years after his breakthrough, Planck still wanted to *confine* his own quantum idea—he drew a line in the physics sand, maintaining that light didn't fly about in little packets, and energy was not atomized, as Einstein suggested.

To return to our school children analogy, Einstein surveys Planck's work and says, "In this strange world, you've made more than a clever mathematical step. I think you've stumbled onto something—the children's heights really *are* exact increments of one foot!" Planck would insist, by stages, that it was some localized circumstance that occurred when the children interacted with yardsticks or when the children misreported their own heights to the teachers. Along with many other physicists, it would take Planck nearly 20 years to truly accept that, indeed, the children of the quantum world only come in heights of three feet, four feet, and so on. Einstein knew it was a radical, even disturbing step. "It was as if the ground had been pulled out from under one," he wrote later.[35]

Although shocking in retrospect, the word "relativity" was nowhere to be found in Einstein's 1921 Nobel Prize, "for his discovery of the law of the photoelectric effect." Relativity was too controversial, mathematical, and theoretical for the Nobel committee, even after its great verification in the eclipse of 1919. Like Planck, Einstein had accumulated an embarrassing number of nominations before receiving the Nobel committee's approval. After the previous Nobel Prize in Physics went to a man who had designed better measuring rods, an Einstein-sympathetic physicist named Carl Wilhelm Oseen joined the Nobel committee. He quickly grasped the committee's anti-mathematical vibe and never mentioned Einstein's relativity work. Instead he wrote persuasively about the link between Einstein's photoelectric effect "law"—he avoided the word "theory"—and successful laboratory results.[36] Since the photoelectric effect could be observed with easy measurements, since it was useful, and since Einstein only needed a bit of algebra to correlate the color of incoming light and the energy of the liberated electrons, this made for award-worthy work.

Even when Planck later came to accept Einstein's hypothesis, and the existence of photons were confirmed in the early 1920s, he still had to keep his own diction: *h* would always be a quantum of action, instead of

Einstein's quantum of energy. "My futile attempts to fit the elementary quantum of action somehow into the classical theory continued for a number of years, and they cost me a great deal of effort," he wrote later. "Many of my colleagues saw in this something bordering on a tragedy."[37]

Some authors have attempted to discern exactly what Planck was thinking in 1900, and a few, including the science historian Thomas Kuhn, doubt whether Planck even deserves credit for the birth of quantum theory. And although Einstein never wavered from crediting Planck exclusively for quantum theory, some wonder if Einstein himself might deserve the greatest share of credit for the quantum era.[38] Still others argue that asking what Planck knew of true quanta presents a question without meaning for 1900.[39] The more appropriate journey might be, as Clayton Gearheart puts it, "to understand Planck on his own terms." Gearheart has presented a thorough overview of Planck's utterances and all subsequent analyses of his thoughts, but in the end, he notes a striking silence from Planck on interpreting his "energy elements."[40] He was clearly uncomfortable crawling onto a shaky limb of conjecture, and he stated plainly (as in his 1906 book on the topic), that his work left a dire need for physical interpretation. In the end, those who credit him with full understanding in 1900, as many physics textbooks do, are oversimplifying the real story of a discovery, and those who would claim he possessed no comprehension of his innovation also paint with too crude a brush. In this sense, the lucid analysis set out by the science historian Martin Klein in 1962 still rings true.[41] Planck knew he needed energy elements and the combinatorics that came with them to correctly derive the new black-body radiation law, and he knew this was a step that, at best, would never fit comfortably within the established physics that came before. Physicist and former Planck student James Franck recalled a Planck talk within a few years of his discovery. "He spoke of some of his attempts to avoid the quantum hypothesis if possible. His conclusions, however, were the following: There is no way out . . . we shall see that it will penetrate into more and more fields of our physics."[42]

Planck's evolution on quantum theory moved slowly—he was usually very cautious when it came to the meaning of his *h*, at least in public and in print. One of the only surviving versions of him thinking off the record is preserved in the letter diaries he maintained with his college friends. In early 1908, he responded directly to a question from Carl Runge.

> Regarding Carl's question about my ideas regarding the elementary quantum, I must first confess that at present they are still pretty poor, but I would

> like to say the following.... the elementary quantum['s] full explanation cannot be done by considering a *state*, but only by considering a *process*. In other words: We are not concerned here with a theory of atomism in space, but with an atomism in time, via processes we usually think of as continuous in time, that have, in fact, temporal discontinuities. That the laws of ordinary mechanics and electrodynamics, which always presuppose temporal continuity, are inadequate here, may well be considered certain.[43]

In other words, while professing public caution for many years to come, the more private and internal Planck considered quantizing *time itself*. He even proposed, in the same 1908 entry, that this might relate to Einstein and Minkowski's spacetime. As of 2014, quantized spacetime has not been proved, but it plays a key role in the efforts to unite Einstein's relativity with quantum theory. Quantized spacetime would mean the fabric of the cosmos has granularity, a grit, as opposed to being absolutely smooth and continuous. In 1908, Planck's notion was many years ahead of anyone else proposing the idea formally.[44] The diary entry also underscores the extent to which Planck knew what was up: He had served quantum poison to "ordinary mechanics and electrodynamics," so classical physics was as good as dead.

Hendrik Lorentz probably made the first dent in Planck's public reluctance to discuss the end of classical physics, even where Einstein had failed. This came just two months after the 1908 diary entry, amidst a public talk delivered in Rome. There, Lorentz made it clear that Planck's law didn't belong with classical theory. Another equation showed the best link to classical ideas; it was the so-called Rayleigh-Jeans law. It completely failed to fit the data, but its derivation was more legitimate and traditional than Planck's, and Lorentz said so.[45] He considered Planck's success "exceedingly curious." By then, several authors had found gaps in Planck's attempts to explain his equation: Lorentz, Jeans, Ehrenfest, and Einstein had all noted quirky, flawed steps as Planck contorted to make his equation fit with traditional physics. Or as Lord Rayleigh stated, some found the derivation impossible to follow.[46]

Planck replied to Lorentz's Rome talk and made his first written concession that something funny was happening with the resonators. When they absorbed or emitted light, his strange new constant h reared up and quantized the exchange of energy between matter and light. But he refused to see light itself as quantized. The 1908 Lorentz Rome lecture was personally crushing to Planck. He heard an international authority—someone he saw as a Clausius-like figure, a nearly flawless elder—cast doubt on his work. And the criticism emerged when the Nobel selection committee was finalizing their options for the year. They initially voted, privately, for Planck to win the Physics prize.[47] Rumors circulated throughout Berlin and even in the popular press that Planck would soon be recognized for his 1900 discovery. Although his equation still fit the data perfectly, most physicists failed to agree on its meaning, and the Nobel Prize committee got nervous enough to reconsider. The 1908 Nobel Prize in Physics went instead to Gabriel Lipmann for discovering color photography.

Meanwhile, Einstein was just as frustrated. Only a few younger physicists embraced his notion of bundled light. His friend and colleague Paul Ehrenfest accepted the truth as early as 1906. And in the summer of that year, Einstein received a letter from Max Laue (who became von Laue in 1913). He apologized to the young Einstein for a slow response, complaining of military service and other obligations. But Laue then wrote that with respect to quantized light, "I have no objections to make." But wasn't he Planck's own assistant? "I have never discussed [this] with my boss. It is possible that there are differences of opinion between him and me on this question."[48]

Outside of Ehrenfest and Laue, Einstein found few sympathizers. Undeterred, as always, Einstein used quantum ideas to stunning effect again in 1907, writing a groundbreaking paper on thermal physics at very cold temperatures.[49] As new cooling techniques opened ever lower temperatures to laboratory exploration, certain material properties changed sharply. Einstein solved one of the open riddles (the strange drop in a material's so-called specific heat value), using the granular energies of quantum theory. This work raised a few extra eyebrows.

In 1908, Lorentz began to accept that something was truly amiss with classical notions of light and energy, just in time for his Rome talk. Willy Wien and Planck began to join them circa 1909, to a limited extent. And in 1910, a few others changed their minds, including the chemist Walther Nernst: "I consider the quantum theory certain."[50]

The watershed moment for quantum theory's childhood arrived in 1911. The Belgian chemist (turned businessman, turned philanthropist), Ernest Solvay underwrote the first conference dedicated to "The Theory of Radiation and the Quanta." This and the subsequent quantum theory meetings were known as the Solvay Conferences. Gathering at the Hotel Metropole in Brussels, 18 leading physicists sifted through the broken shards of classical theories of energy and light. Einstein, at 32, was the youngest, and he began the conference with a clear call to arms, starting with his updated work on specific heats. But the old guard bombarded him with challenges. Poincaré, Lorentz, and Planck led the way in voicing doubts.[51] They didn't want to toss out the good classical baby of electromagnetic waves with the quantum bathwater. Einstein found the experience frustrating and even depressing. He wrote to a friend that the meeting felt to him like, "a lamentation on the ruins of Jerusalem."[52] And about Planck in particular, Einstein wrote that the esteemed professor was "stubbornly attached to preconceived opinions that are undoubtedly false."[53]

As Planck wrote afterward, older, established scientists were duty-bound to convey "an increased caution and reticence in entering into new paths."[54] From his point of view, he had already shifted his thinking a great deal. He found the meeting extremely tiring, but he would look back on that week, with conversations varying from exasperated to breathless, as one of the peaks in a long career.[55]

A year later, Planck's second edition of *The Theory of Heat Radiation* gave an interesting window into the awkward childhood of quantum theory and his ongoing worries. In the preface he wrote that his ideas on black-body radiation "have met with little general acceptance." He painted himself as the reasonable pioneer caught between two camps: those who, "through conservatism, reject the ideas," and a few others who, "have felt compelled to supplement [my ideas] by assumptions of a still more radical nature, for example, by the assumption that any radiant energy whatever, even though it travels freely in a vacuum, consists of indivisible quanta or cells." He admitted that his first edition, from 1906, was deficient. But he rejected the notion that such elements were relevant to anything but the interface between matter and light; Einstein went too far for Planck, even well after the Solvay conference. "Since nothing probably is a greater drawback to the successful development of a new hypothesis than overstepping its boundaries, I have always stood for making as close a connection between the hypothesis of quanta and the classical dynamics as possible."[56]

So in 1944, when Planck's remaining colleagues crawled through rubble, brushed the dust from their tuxedos, and insisted on toasting his career, they celebrated a story of Max the way-shower. They lauded a revolutionary who brought quantum to the world, even if he had only reluctantly played along after his first inspiration. Certainly, during the speeches of that evening, they noted his prolific ways. In a scientific era predating our current obsession with fire-hose–like research output, Planck published an array of books, 42 notable papers in *Annalen der Physik* spanning six decades of work, and just as many articles in other outlets. They would have mentioned the impressive expanse of his physics gaze, with key publications as far flung as thermal physics, fluid dynamics, and relativity theory.[57] His colleagues might have mentioned his philosophical battles and victories versus the positivists, those who would constrain human imagination. The assembled had to toast his great successes as a leader and administrator of German physics and a spokesman for German science in the face of absurd challenges. But in the end, the crowning achievement of their Geheimrat Professor Planck would always be conceiving quantum theory. After accumulating 75 Nobel nominations over more than 20 years, he finally won the 1918 prize in 1919, for "his discovery of the energy quanta."[58]

Looking back from a century's remove, we can try to set out Planck's most lasting contributions. His igniting quantum theory leads the list, but it would be a sad mistake to stop there. He also elevated the concept of entropy, gave the world the notion of natural units, and discovered the ubiquitous "zero-point" energy.

Planck's focus on entropy (and the closely associated idea of irreversibility) was far ahead of its time. It is no exaggeration to say he took a rough hewn entropy notion from Clausius, like a shapeless piece of flint, and carved it not just into a useful blade but a Swiss-army knife. Thanks largely to Planck, entropy became quantifiable, statistical, and useful. Today, an enormous range of scientific studies employ entropy as a tool, including something as extreme as black hole thermodynamics, where the size of the hole's boundary makes a nice analogy to its entropy,[59] or something as practical as information theory, where the basic measure of information is its "Shannon entropy." And in repositioning entropy in the second law of thermodynamics, suggesting that it always tends to increase, Planck underlined the so-called arrow of time. That the universe moves irreversibly in one direction is accepted, yet still mysterious. Some leading physicists suggest we might be well served by focusing more of our attention here, despite thousands of years of head-scratching.[60]

Although Planck was humble about his early thermodynamics work, his contributions here should not be overlooked. Science historian Dieter Hoffmann argues that Planck, stripped of his black-body work, would still be famous today for his pioneering work in thermodynamics. If one takes citations as the ultimate measure of a scientific author's impact, the data underscore Hoffmann's claim. The two most-often cited of Planck's papers come from his early work, linking entropy to the behavior of chemical solutions.[61] And Planck had kept contributing to thermodynamics. In 1911, the same year as the Solvay conference, he had put the third law of thermodynamics into its most palatable form: As the temperature of an object approaches absolute zero, so too will its value of entropy.

Planck also suggested natural units in the last breaths of the nineteenth century. And when he then gave the world *h* and *k*, he recognized an extraordinary opportunity. "With their help we have the possibility of establishing units of length, time, mass and temperature, which necessarily retain their significance for all cultures, even unearthly and nonhuman ones."[62] Every universal constant has units. The speed of light *c*, for instance, sensibly measures distance (meters) divided by time (seconds). In the case of *h*, the units are energy multiplied by time. It's not always pretty—the universal constant of gravitation, *G*, has units of distance cubed, divided by both mass and time squared. But combinations of the various fundamental constants can provide absolute measures of time, space, energy, and so forth. Hence we have, for instance, the Planck time and the Planck length—they give a voice to the universe in its own terms. Independent of human bias, they have proved a boon to the further development of theoretical physics, in terms of both new perspectives and also the subsequent versions of the natural units. (They provide a kind of style manual for all physicists who do their work with pencil, paper, and laptop.)

And finally, any list of Planck's contributions would be incomplete without the mysterious notion of a zero-point energy. In his second approach to deriving his radiation law, his slightly modified mathematics meant that, even if an object had a temperature of absolute zero, there would still be some sort of *energy* remaining in it. Though that seemed an oddity at the time, a necessary but inconvenient mathematical bit of detritus, the idea has only grown in importance over the last 100 years. The now-accepted idea that space itself has a seething and invisible lake of energy in it, even without particles or light involved, has implications for the structure of the universe itself.[63] The zero-point energy of the cosmos is automatically a candidate for determining the source of "dark energy" that appears to push the boundaries of the universe outward with ever greater speed. The dark energy is still mysterious as of this writing, and it is not properly accounted for by any computation of zero-point energy, but the two often appear in the same conversations. And since any discussion of dark energy is tied to an idea that Einstein originally launched as a "cosmological constant," it is accurate to say that Einstein and Planck debate one another intensely even today, as quantum theory and general relativity still confront one another. Like their authors once did, the two theories may even need one another, despite the appearance of irreconcilable differences.

In a joint 1932 interview, both men underlined an irrational faith in the universe as a scientific necessity. "As Einstein has said, you could not be a scientist if you did not know that the external world existed in reality," Planck said. "But that knowledge is not gained by any process of reasoning. It is a direct perception and therefore in its nature akin to what we call Faith. It is a metaphysical belief."[64] In that interview, as an exemplar of this Faith, Planck noted the scientist Johannes Kepler—300 years his senior—with a tone of ultimate respect, even awe.

> He was always hard up. He had to suffer disillusion after disillusion and even had to beg for the payment of the arrears of his salary by the Reichstag in Regensburg. He had to undergo the agony of having to defend his own mother against a public indictment of witchcraft. But . . . what rendered him so energetic and tireless and productive was the profound faith he had in his own science, not the belief that he could eventually arrive at an arithmetical synthesis of his astronomical observations, but rather the profound faith in the existence of a definite plan behind the whole of creation. It was because he believed in that plan that his labour was felt by him to be worth while and . . . by never allowing his faith to lag, his work enlivened and enlightened his dreary life.[65]

Kepler makes a compelling analogue for Planck, and not just as Germans or even as scientists overcoming incredible personal hardship. Working in the seventeenth century, Kepler took a set of data that defied all reasonable description—the detailed nightly positions of the planets in the sky—and found the perfect equations to describe these measurements. When looking at enormous compilations of planetary positions, he began to see a mathematical pattern and with his new laws of planetary motion, he solved a fundamental riddle: the structure of the solar system. Kepler's laws of planetary motion apply equally well to other star systems as well; they are as universal as Planck's law of thermal radiation. After trying several ill-fated mathematical approaches to explain an exhaustive list of planetary positions, he finally began to see ellipses in them. Mathematics had always been Kepler's chief tool; childhood smallpox had left his hands and eyes deficient for using telescopes or lab instruments.[66] In a real sense then, as Kepler didn't collect measurements himself, he was a true predecessor of Planck and one of the first mathematical physicists. Just as Kepler had an astonishing ability to see the mathematical form in confounding planetary data, Planck had the same when looking at the black-body radiation curves. While Kepler would not live to see his name immortalized, Planck did live to enjoy recognition and the toasts of colleagues.

For a night so bomb-free that Heisenberg called it "spooky," surviving Berlin colleagues recalled the superlative decades of their Max Planck. He held an almost religious significance for them. According to the journalist James Murphy, whether you spoke with Planck about science or anything else, "one often feels that this tragedy of his children has made a deep impress on his soul. The memory of it seems to evoke a certain wistful quality which is profound in his nature and give it the warmer glow that one is inclined to call mystic."[67]

This mystic of German science would have endured his painful back with unusual ease that night. His spirits lifted with so many friends and memories so close. As the guests dispersed, he would have stood with Erwin. Did they embrace as they said their good nights? Max probably suggested another Rogätz meet up, whenever Erwin and Nelly could visit again. But the banquet's end marked the last time Max would see and touch his son before the Gestapo stepped between them.

11

July 1944—An Arrest

On July 10, Erwin wrote his father a letter. "I hope you recovered from the party in Berlin." And he promised to visit Rogätz later in the month. Max replied eagerly, welcoming a possible visit. "The strawberries are about done," he wrote. "But the cherries are starting."[1]

The two had a long history of letters, starting from father to child, then from war prisoner to father, and finally between closest friends. Among their first were a pair of notes when Erwin was five years old. He wrote a riddle to his father, and Max wrote back, in the manner of what we might today label a goofball.

"Dear Erwin, You wrote me a very difficult riddle. I was thinking for a long time which type of boots you cannot wear. Maybe the heaven boots, or the dirty boots. Stop! Now I know. The lost boots—those you can most certainly not wear. Your faithful father."[2]

In Erwin's childhood, they would exchange letters either when Max traveled for his work or when Marie and the children summered at Tegernsee. In 1900, the year of Planck's great breakthrough, the letters show him testing Erwin's spelling. He felt his son needed to improve, so he inserted errors and made Erwin find them. As he signed one such note, "From your faithful father (altogether six mistakes)."[3]

The playful Max reveals himself again in a 1901 letter, as the children came to join him at a scenic retreat. "Dear Erwin, Because you turn 8 today, you shall get a present, a nice journey to Rudolstadt, but since you are still too little to take the train all by yourself, you can choose who will go with you. You're allowed to take two girls and an older boy. Now choose who shall go with you, and tell me, so I can buy the tickets. Your father."[4]

In his twelfth year, Erwin became gravely ill with appendicitis, and after a successful surgery and recovery, Max allowed a special treat: a trip for Erwin and his mother riding in a *second-class* train car, the type with seat cushions. This year also marked Erwin joining the family's chamber orchestra, as he received a cello for the previous Christmas.[5] And after World War I, he played the cello in a series of trios with his father and Einstein.

The physicist and his political son comforted one another in the aftermath of their mutual tragedies. And over the subsequent two decades, their dinners, concerts, vacations, and ambitious mountain treks dotted the yearly calendars (Figure 11.1).

Figure 11.1. Max Planck with son Erwin in the mountains in the late 1930s. Courtesy Archiv der Max-Planck-Gesellschaft, Berlin-Dahlem.

As Max awaited a visit from his son in 1944, Erwin's political colleagues triggered a desperate plan. On July 20, 1944, Adolf Hitler entered a mid-day meeting at his eastern "wolf's lair" headquarters (in what is today northeastern Poland). About 20 of his top military officers circled a table to discuss the war; they would stack bad news before the Führer, as the Russians advanced from the East, and the British redoubled their bombing. Shortly after the meeting began, Colonel von Stauffenberg left the room to take a phone call, while his bomb-loaded briefcase stayed under the table. Moments later, a tremendous explosion tore through the room and its inhabitants. Max Planck would have heard about it through the German News Agency later that day: "The German people must consider the failure of the attempt on Hitler's life as a sign that Hitler will complete his tasks under the protection of a divine power."[6]

Although four died in the blast, the singed Führer survived with a concussion and an injured arm. But he even kept a meeting with Benito Mussolini in the afternoon. He gave a radio address to the Fatherland the next day, praising fate as the Gestapo quickly nabbed the conspirators, mainly members of the military. Hangings commenced at the Plötzensee Prison in Berlin.[7] Since Erwin would have immediately heard reports of the failed attempt, why didn't he flee? He had a valid passport (for his international work), and he must have known he'd be a person of interest. Other conspirators later said they had pledged to wait out the days after the bombing, come what may.[8] Erwin may have doubted the reports and hoped to still join a post-Hitler government. Perhaps he simply couldn't stand the thought of abandoning his father.

On July 23, the Gestapo cast a larger net, with Heinrich Himmler's agents arresting more than 5,000 in loose connection with the bomb plot.[9] They took Erwin from Nelly and their home with no explanation. Others arrested included children of the von Harnack and Delbrück families, playmates of Erwin's suburban youth.[10] (One of the children, the eventual Nobel laureate Max Delbrück, had fortunately taken his biophysics career to America, but another famous son of Grunewald, the theologian Dietrich Bonhöffer, was already in prison; the Gestapo had caught him helping Jews escape to Switzerland in 1943.) Erwin was charged with aiding the conspirators, especially through his friendships at the so-called Wednesday Society. This group of intellectuals, like some sort of earnest and exclusive book club, had long gathered to take turns hosting dinner and composing essays for the assembled. General Ludwig Beck, a leader of the July plot, was known to be a vital member there. Max Planck had attended meetings in the past, but not with Erwin's frequency during the Nazi years.[11]

The Gestapo denied Max and Nelly's requests to visit Erwin—the charges against him were too serious. In custody, he was brutally interrogated and tortured.[12] He admitted that he had met many of the conspirators years earlier. In the late 1930s, he said they had discussed how to avert or end the war using their diplomatic channels.[13] He claimed no knowledge of the assassination attempt.

But in fact, Erwin was a known opponent of the Nazis and had been active in the German resistance for years.[14] Specifically, he penned sections of a draft constitution as early as 1940, preparing for a post-Nazi government. And Erwin's name appeared prominently on a "critical personnel" list maintained by one of the briefcase conspirators.[15] Although a conservative politician, strongly aligned with military hawks, Erwin and his closest allies had opposed the NSDAP (the Nazi party) from its earliest times.

After Hitler assumed power in early 1933, Erwin thought it best to be scarce, even if it meant leaving his dear Nelly behind. He departed that spring on a long trip through Southeast Asia, visiting a series of German diplomats along the way.[16] The Nazis consolidated power and created a totalitarian state within a few months. One year later, during a violent purge known as the Night of the Long Knives, Hitler's agents burst into General Kurt von Schleicher's home, executing both the general and his wife. The purge both tamed disorderly Nazi factions like the *Sturmabteilung* (the Nazi paramilitary group known by their brown shirts), and also eliminated perceived political rivals, like Erwin's friend and former boss. The military mostly approved of the purge, even as it lost some of its high-ranking leaders, since they loathed the *Sturmabteilung*. When Erwin learned of von Schleicher's death, he immediately wanted answers—how was this possible without a trial, and how could his old colleagues stand by quietly—but he found only intimidation, as Nazis confiscated wreaths and flowers from the funeral. According to Erwin's forced 1944 testimony, his grief and anger in 1934 led him to his first anti-Nazi activity, as he tried to spark sympathetic army officers to take action against them.[17] "If you do not lift a finger," he told one general, "you will meet the same fate sooner or later."[18]

In 1936, Erwin went to work for the Otto Wolff Company, a steelmaker and one of Germany's largest firms. He employed his diplomatic skills to construct and refine foreign contracts, at first working from Cologne. (Despite his opposition to the Nazis, Erwin's work helped feed raw materials into the Reich's war machine.) To his father's delight, he returned to the company's Berlin office in 1939, shortly after the outbreak of war.[19]

What Max Planck knew of his son's involvement with factions of German resistance is open to speculation. He maintained shock at Erwin's arrest and declared his son's innocence, but he was also incredibly sharp, even at age 86. While Erwin apparently didn't share his secret political activity with his father, the two were very close and had talked politics for years. We don't know what Max may have seen and heard in Nazi-era visits to the Wednesday Society.

His own politics leaned conservative, matching his upper-class Prussian upbringing. But his family was never in the most nationalist camps. As with so many facets of Max Planck, he traced a path of moderation, seeking the middle ground. He can be contrasted, for instance, with his lifelong friend Willy Wien. They had met in their 20s during an East-Prussian hunting retreat. While the Wien family had supported Otto von Bismarck's ambitions to expand their fledgling empire, the Plancks disagreed.[20] The Wien clan was also strongly anti-English, a stance that would only lock its knees for Willy after World War I. More than anything, Planck's political arc rises to one severe 1914 public wound followed by a lifelong retreat and a willful public distance best summarized by his distaste for, "the fundamentally unnatural mixture of scientific and political activity."[21]

As he became a more established professor through the 1890s, he began to offer blunt political opinions where they affected the academy. In 1895, two cases had Planck standing up to the Kaiser's government, refuting its interference in normal university business. First, although he and his colleagues recommended Emil Warburg as the best candidate for a professor position, the Prussian education minister opted to consider other finalists—he didn't rubber stamp the Berlin faculty's opinion. Planck protested, standing up for the young Jewish physicist: "One need not be pro-Semitic to find such a procedure highly questionable." Further, he wrote to the minister that this meddling "disregards ... the authority of the faculty, which reached its conclusions after careful deliberations."[22] In the end, Planck won the day and the university hired Warburg.

Later that year, and for several years to follow, Planck waded into the thorny case of a Jewish physics lecturer named Leo Arons. Aside from his entry-level physics post, Arons was independently wealthy and supported left-leaning causes, like underwriting a leftist newspaper. The Kaiser's culture minister demanded that the University of Berlin discipline Arons for crass political activity in defiance of the empire. In response, a special academic commission formed to review Arons's work and respond accordingly: Planck, along with the history professors Theodor Mommsen (still famous today for his history of Rome), and Heinrich von Treitschke. Planck's commission found Arons's work praiseworthy, and they refused to discipline him. Even after the government decreed that it could discipline faculty directly, the university and the commission held their ground.[23] They defended their liberal, Jewish colleague (even though Planck was center-right, and von Treitschke was by that time an outspoken anti-Semite).[24] Mommsen and Planck stand together again today, two adjacent statues on the university's *Unter den Linden* campus.

Increasingly comfortable in political settings, Planck ran heedlessly into his greatest shaming during the earliest days of World War I, with the empire consumed by a prideful fever. In a bid to outflank French troops and march on Paris, German armies moved quickly through Belgium first. Reports of looting and crimes against civilians emerged in short order. But for the German families at home, the idea that their young soldiers displayed poor etiquette, never mind criminal activity, was unimaginable, and coming from non-German sources, the stories were viscerally insulting. Erwin Planck, among the earliest Western deployments, must have trooped through Belgium as the reports surfaced. So Max Planck was presumably as dismayed reading the allegations as any other German father.

Feeling the need for a quick response, separate from the army or the Kaiser, a band of prominent academics and artists issued a joint decree labeled an "Appeal to the Cultured Peoples of the World," and henceforth called the Appeal of the 93 Intellectuals. Largely written by German playwright Ludwig Fulda and published before many signatories (including Planck) had read it, the appeal supported German military acts as righteous.[25] "As representatives of German science and art, we hereby protest to the civilized world against the lies and calumnies with which our enemies are endeavoring to stain the honor of Germany in her hard struggle for existence—in a struggle which has been forced upon her." It went on to specifically deny that the Kaiser wanted the war, that the army trespassed in neutral Belgium, or that the "the life and property of a single Belgian citizen was injured." The 93 concluded, "We cannot wrest the poisonous weapon—the lie—out of the hands of our enemies. All we can do is to proclaim to all the world that our enemies are giving false witness against us."[26]

Joining Planck with their notable signatures were the physicists Philipp Lenard and Willy Wien, chemists Walther Nernst and Fritz Haber, biologists Ernst Haeckel and Paul Ehrlich, and the mathematician Felix Klein, as well as Planck's otherwise-rational neighbor Adolf von Harnack. Einstein refused to sign and even attempted to start his own joint letter preaching against the insanity of war. Max was the only one of the 93 Einstein approached about this alternate letter, but he declined to sign and Einstein gave up.

In truth, the German army killed thousands of Belgians and burned thousands of homes. As if these heinous acts were not already counter to "intellectuals," the German army torched the University of Leuven's library, including hundreds of thousands of medieval documents. The Entente powers labeled late August 1914 "The Rape of Belgium," and often added exaggerations to help rally their own populations to the war.[27]

Planck was shown his rash error by Hendrik Lorentz, by then the kindly papa bear of physics—trusted, heeded, and respected by all parties. In a series of patient letters, Lorentz brought Planck to understand the reality of some Belgian stories and to revise his stance. When, in late 1914, Willy Wien wrote a new manifesto, decrying the ethics and trustworthiness of British scientists, Max declined to join. (Johannes Stark and 14 other physicists did sign it.) As Planck wrote to Lorentz in March of 1915, "scientists face no more urgent or finer challenge than to do their best quietly to counter . . . the deepening of hatred among peoples."[28] In early 1916, he published an open letter to Lorentz, but only after Lorentz had reviewed a first draft, buffing the roughest rhetorical edges. In the letter, Planck noted "with distress" that the original Appeal generated "incorrect ideas about the feelings of its signers." Furthermore, he would "not defend the behavior of every German, either in peace or in war." He closed by expressing his conviction "that there are domains of intellectual and moral life that live beyond the struggles of nations, and that honorable cooperation in the cultivation of these international cultural values and, not less, personal respect for citizens of enemy states are indeed compatible with ardent love and energetic work for one's own country."[29] Planck encouraged Lorentz to share this widely, and particularly with the esteemed multinational group who had attended the first Solvay conference.

Planck's stance may have softened with his son a prisoner—the French captured Erwin shortly after the Appeal of 93 Intellectuals. But he was also witnessing the war's shocking costs. His colleague Nernst lost two sons during 1915, and Planck himself lost a nephew. When he wrote the foregoing letter, Planck's immediate family was still intact, but just two months later, the fighting near Verdun claimed Karl Planck.

The Appeal clouded Germany's post-war scientific relationships for years to come. The International Research Council (IRC) formed in Britain after the war and supplanted the Germany-based International Association of Academies. As a first act, the IRC forbade scientists of former Central Power nations from administrative posts and meetings. Even Planck, devoted to restoring international ties, bristled at the IRC. He wrote to Lorentz that he and his colleagues preferred a more purely scientific body (like the former organization), versus the IRC, "in which the political element naturally plays a much greater part."[30]

For the years immediately following the war, Einstein was a lone emissary for German science. His new fame following the 1919 eclipse measurement, combined with his refusal to sign the Appeal of 93 Intellectuals, landed him a series of invitations, even as resentment grew at home. One imagines the scientist shaking his head at the term "home"—he never embraced the concept, in Germany or elsewhere. Planck's softening attitude in the war's first year and his embarrassment over the Appeal of the 93 nurtured Einstein's growing esteem for Planck. "He is a splendid fellow," Einstein wrote to a friend in 1915.[31] Whatever fondness grew between them, it was never based on politics. Even in late 1917, after the loss of Karl, Planck underlined his views in a letter to Einstein. " 'Sound progress' and 'conservative' are not only not contradictions, but inseparably bound together." Planck cites an improving military outlook and a strong exchange rate. And as he wishes a digestion-plagued Einstein better health for the year ahead, he adds, "I hope that during the same year your sympathies for the German Party grow as well."[32]

During the Weimar years, Planck belonged to the right-of-center Deutsche Volkspartei (the German People's Party), friendly to industrial concerns and misty-eyed for the days of monarchy. While the party's platform included notions of equality for women and labor rights, it also supported an aggressive military stance. It never embraced the Weimar Republic and at times openly rejected the notions of democracy. Planck himself was much more a monarchist than a modern liberal. Ten years after Hitler took power, Planck condensed the rise of the Third Reich to friend von Laue. "The fundamental evil consists, in my opinion, in the coming of the dominance of the masses. Indeed, I believe the general right to vote (for twenty-year-olds!) to be a fundamental error."[33]

Throughout the Weimar years, Planck restricted his political activity to that which affected science directly: He advocated for funding and helped create new financial organisms to keep German science afloat. And he fought the misinformation spewing from Lenard and Stark's *Deutsche Physik* movement. Even in these cases, his work was more behind the curtains than at the podium. At the same time, he prepared to smoothly wind down his university duties. He officially retired from his professorship in 1926 but maintained most of his responsibilities, including lecturing, for another year as he awaited his eventual replacement, Erwin Schrödinger.[34]

Two indirect political acts of the time demonstrate either Planck's political tin ear or his commitment to holding politics far from his nose. In 1925, he happily accepted an invitation to deliver a keynote address in Russia. To a group of 1,500, he said, "There is no bond that unites the different countries of the earth, with their widely divergent interests, so directly and harmoniously as science." Certain parties within Germany decried his fraternization with Bolsheviks, and Planck's visit had indeed conferred an increment of legitimacy on the Soviet Union. He simply noted that the festivities closed with Beethoven's Ninth Symphony, as opposed to any sort of political anthem.[35] Just a year later, he visited the opposite end of the spectrum for the 100th anniversary of Alessandro Volta's death. In this case, Planck joined the celebration in fascist Italy, later noting that dictator Benito Mussolini was well behaved and avoided making any sort of ugly political statements.[36]

Approaching his seventieth birthday in 1928, Planck was well into retirement, but he watched a sad constriction of his social circle. Willy Wien passed in August, just a year removed from the death of mathematician Carl Runge, another lifelong friend. But Planck nurtured an active array of correspondence on science and philosophy. He was particularly united with Einstein in their distaste for the emergent views of quantum mechanics. He and Marga enjoyed relaxing vacations in these years, with mountain climbing and afternoons of lakeside reading in the Alps. So too were his hours with Erwin especially precious, though his son was working feverishly within the teetering Weimar government.

Einstein sought increasing distance from Berlin's political and racial strife. By the end of 1929, he completed his getaway cottage in Caputh, 15 miles southwest of Berlin and not too far from Grunewald. The house perched on terraced slopes and featured a comfortable deck overlooking a lake, as well as an observing platform with a telescope.[37] One imagines Einstein alone at night probing the cosmos as Galileo had done three centuries earlier. A major impetus for Einstein's retreat from Berlin was the rising anti-Semitism there. In a 1929 interview with the *Saturday Evening Post*, he was asked if Jews should try to assimilate. "We Jews have been too eager to sacrifice our idiosyncrasies in order to conform." On the balance beam of being both German and Jewish, he said, "Nationalism is an infantile disease, the measles of mankind."[38] This contrasts with Planck's own statements of the era, describing how the best science blooms. "The history of international science has shown again and again that science, just as art and religion, can prosper in the first instance only on national soil."[39]

Even so, beneath politics the two shared an incredible connection. In a 1933 joint interview with Planck, Einstein compared humanity to "a juvenile learner at the piano, just relating one note to that which immediately precedes or follows . . . it will not do for an interpretation of a Bach Fugue."[40] The assumption that the universe has a composer and a beauty, apart from our existence or understanding, is never in doubt. In the same interview, Planck shared his joy at our permanent ignorance. "And if we did not have faith but could solve every puzzle in life by an application of the human reason what an unbearable burden life would be."[41] By the time their eloquent duet hit bookshelves, their friendship would effectively be broken by the rise of Nazism.

The kindling of nationalism was soon to be lit by economic calamity. Warnings of impending crisis were not hard to find. If Germany possessed a hero during the Weimar era, it may have been Gustav Stresemann, the Republic's chancellor and then foreign minister. Hailing from Planck's preferred Deutsche Volkspartei, he deftly wove his way through mazes of German and international politics to preserve the fragile peace of the 1920s. He renegotiated Germany's war debt to one fourth its original size, and he is largely credited with solving the nation's credit crisis and ending the hyperinflation of 1923.[42] Stresemann had won the Nobel Peace prize in 1926 for his role in several intricate treaties keeping Germany both afloat and out of renewed European conflict. But shortly before he died in 1929, he famously said, "We have been living on borrowed money. Should a depression occur and the Americans withdraw their short-term credit, then we will be bankrupt. . . . We have no means."[43]

Weeks later, after the American markets swooned in their Black Friday, Germany's life's blood of U.S. loans stopped cold. Then American markets, via trade tariffs, closed themselves to fragile German industries. The weak Weimar government could no longer pay its war loans and German unemployment soared again, with one in four Germans being out of work by Christmas of 1929.

Planck tried to relish his retirement but couldn't avoid the renewed troubles. To celebrate his seventy-second birthday in April of 1930, Planck climbed snow-packed Jungfrau, the third-highest peak in the Bernese Alps.[44] If he cleared his head and (briefly) enjoyed the view, he then descended into a surprising maelstrom of new work. In May, Planck's friend Adolf von Harnack, the first and founding president of the Kaiser Wilhelm Society, stepped down in failing health, and he passed away in June. Planck felt duty-bound to take on this mantle, representing German science at the most difficult time. As the daily life of Germans worsened, their interest in science plummeted. Many began to see modern technology as a job destroyer. Simultaneously, their flirtation with extreme ideologies moved from casual glances across the bar to hope-filled conversation. Between 1928 and 1932, Germany's unemployment grew nearly 10-fold, and the Notgemeinschaft's budget for physics was cut by more than 50%.[45] In the same period, with Hitler promising a new job for every man, the NSDAP went from polling less than 5% to polling well over 30% in Reichstag elections. As job openings evaporated, campus students watched their ranks swell with disgruntled job seekers, and they embraced the NSDAP even more quickly than their parents. The Nazis polled about 10% of students in 1929, but more than half declared Nazi sympathies by 1931.[46]

Einstein wrote to his friend Planck in late 1930. As they watched the economic smoke rising from around the world, Max particularly fretted about the rising Soviet menace—he had just helped write a declaration against a purge of Soviet scientists. Einstein said they need not look so far afield. "Even in our case external conditions are developing slowly but steadily in a threatening direction."[47]

With election posters demanding "freedom and bread" in red ink, the Nazi message of hatred for Bolsheviks and Jewish bankers resonated with a populace thirsting equally for security and enemies. When the new Reichstag convened, the Nazis now trooped in the second-largest contingent of representatives. As German banks fell like so many exhausted and unwatered horses, Nazi party membership continued to climb.

And amidst the trouble, Erwin continued his ascent of the rickety Weimar Republic scaffolding, moving in 1932 to his highest position yet, State Secretary under new Chancellor Franz von Papen. Von Papen's highly aristocratic cabinet was routinely ridiculed as the "cabinet of barons" or the "cabinet of monocles."[48] (Erwin didn't wear a monocle, to the best of our knowledge, but was lumped in with this group nonetheless, as a reference to the *Junkers* of a bygone era.) Von Papen moved against the Reichstag, dissolving them just after his appointment and calling for new parliamentary elections in July. The future of the broken Republic was clearly at stake.

Einstein wondered about the coming elections in a June 1932 conversation with journalist James Murphy. At the Caputh lake house, Murphy asked if Einstein might write an introduction to Planck's new essay collection, but Einstein surprised him.[49] "He said that it would be presumptuous on his part to introduce Max Planck to the public," Murphy recalled, "for the discoverer of the quantum theory did not need the reflected light of any lesser luminary to show him off. That was Einstein's attitude toward Planck, expressed with genuine and naïve emphasis." Murphy pressed his case, saying this book would be published for the English-speaking world, where Planck was less known. Einstein, "would have been pleased if the truth were the other way round."[50]

By the end of the year, Einstein had written the introduction. It includes the beautiful description of the Temple of Science, with Planck representing the few who enter for the purest of reasons. In introducing Max Planck to English audiences, Einstein concludes with two sentences, one prescient and the other poignant. "His ideas will be effective as long as physical science lasts. And I hope that the example which his personal life affords will not be less effective with later generations of scientists."[51] These were among the last kind words to pass between them.

By 1932, Einstein's interest in foreign appointments was well advanced. After serious flirtations with the California Institute of Technology and Oxford, a newcomer finally caught his eye: the fledgling Institute for Advanced Study, adjacent to Princeton University. He finalized a contract with the Institute's director in June.[52] To be sure, Einstein had his doubts about America; when his migrant physicist friend Paul Ehrenfest wondered about posts in the United States, Einstein told him not to bother, since

America housed "a boring and barren society."[53] He also suggested his new position was neither full time nor permanent, and he probably hoped to juggle part-time appearances in England, California, New Jersey, and, pending the political situation, Berlin. His new American contract became internationally known in August. "I am not abandoning Germany," he said then. "My permanent home will still be in Berlin." But at the same time, he confided to a friend that he didn't see how Chancellor Franz von Papen and President Paul von Hindenburg could impede, "the imminent National Socialist revolution."[54]

With his second wife Elsa, Einstein gathered 30 pieces of luggage and made his fourth trip to America starting in December of 1932; although he had talked with colleagues of returning the following spring, he rightly understood he might never live in Germany again.[55]

Meanwhile, Erwin Planck found himself in the middle of high-stakes brinksmanship. At the end of 1932, as Chancellor von Schleicher's chief of staff, Erwin would have experienced the intense conversations that led to Hitler's political acceptance. The Weimar government hit a new crisis point in January 1933, with von Schleicher maneuvering for power against Franz von Papen. Tragic in retrospect, von Schleicher had extended a hand to Hitler and the NSDAP to bolster his own fractured base of support. Since von Schleicher doubted the loyalties of his mostly inherited cabinet, Erwin was one of the few he trusted. When it became clear that he was losing the confidence of President von Hindenburg, von Schleicher schemed to keep the chancellor's position away from his archrival von Papen. His short-term fantasy was to have Hindenburg place Hitler as chancellor, with himself as minister of defense and head of the armed forces. "If Hitler wants to establish a dictatorship, the Army will be a dictatorship within the dictatorship" von Schleicher told one associate, and presumably Erwin heard similar comments. Hitler hinted that he would support this structure, but meanwhile continued his own multi-front back-channel scheming. In short order, Hindenburg appointed Hitler as chancellor. But as von Schleicher awaited a new assignment, Hindenburg fielded rumors that the General was plotting a coup. The president dismissed von Schleicher from government service.[56]

With his friend and former boss having lost everything in the shuffle, Erwin Planck resigned his position on the day of Hitler's appointment. He began finalizing plans for his yearlong trip abroad—he probably predicted a bumpy transition to come, especially as a member of the former government.

In March 1933, while in California, Einstein granted an interview to the *New York World Telegram*. "As long as I have any choice in the matter, I shall live only in a country where civil liberty, tolerance, and equality of all citizens before the law prevail. These conditions do not exist in Germany at the present time."[57] Now he and his odd-couple partner Planck would carefully and almost delicately cross swords. Einstein thought he was making necessary statements to awaken the world, whereas Planck thought his old friend was airing laundry out of turn and inflaming a turbulent, but temporary, transition. Yes, the NSDAP was embarrassing and even misguided, but like many Germans he probably assumed they would mature when faced with governance. In any case, it was unthinkable to stand with a megaphone and exaggerate their warts for all the world to hear.

Max wrote to Einstein a week later, emphasizing that the substance of any action lay in its consequences more than its motives. "By your efforts your racial and religious brethren will not get relief from their situation, which is already difficult enough, but rather they will be pressed the more."[58] Meanwhile, the Nazis seized Einstein's German bank account, padlocked his Berlin apartment, and even raided his beloved cottage in Caputh. He then renounced his German citizenship and wrote a resignation letter to the Prussian Academy of Sciences. "I consider my position's inherent dependence upon the Prussian government intolerable."[59]

Planck's next letter to Einstein was achingly pragmatic, stating plainly that the new regime and Jewishness were, "ideologies that cannot coexist," and that he was honor-bound to Germany, no matter the leadership. He noted, "This idea of yours (resignation) seems to be the only way that would ensure for you an honorable severance of your relations with the Academy." But, "despite the deep gulf that divides our political opinions," he wrote, "our personal amicable relations will never undergo any change."[60]

Einstein quickly replied, assuring his friend that he felt the same way: Their friendship would rise above whatever unpleasantness roiled below.[61] But he had to show his deep wound as well. "I was just useful to Germany's reputation . . . no one thought it worth their while to stand up for me."[62] Einstein never returned to Germany, even refusing invitations from his friend von Laue. In fact, his only conciliatory gesture came long after World War II when he began sending nominations for the German Physical Society's annual Max Planck medal. He and Max had been the first two recipients, not long before Einstein left for America.[63]

In the spring of 1933, Max and Marga decided to embark on their planned vacation to Sicily, despite the turmoil. As she wrote to Paul Ehrenfest that month, "He has often wished that he could withdraw from official matters and great responsibility. . . . Now everyone counts on his help."[64] Much would happen in Berlin during their absence, and he would soon receive worried missives from colleagues, with von Laue begging him to abandon his trip. But Planck refused and even predicted "all the troubles will be gone by the time I return."[65]

Meanwhile, most German scientists rallied against Einstein. The Prussian Academy of Sciences issued an April 1 press release approved in a nearly unanimous vote. The press release claimed the Academy was, "shocked to learn from newspaper reports about Albert Einstein's participation in the loathsome campaign in America and France. . . . Einstein's agitatorial behavior abroad is particularly offensive to the Prussian Academy of Sciences. . . . For this reason it has no cause to regret Einstein's resignation."[66] Even Fritz Haber, Einstein's friend but a Jew dedicated to assimilation, voted in support. The stubbornly vacationing Planck never saw the draft and missed the meeting. For Max von Laue, the vote was, "one of the most appalling experiences of my life."[67] Stark and Lenard both crowed for the historical record. Stark wrote an open letter to *Nature* magazine about Einstein's deplorable behavior and how it would negatively affect German Jews.[68] Stark had become increasingly marginalized in Weimar Germany, with his critics nicknaming him "Robustus" and "the Starkest man" behind his back.[69] With a new regime in charge, he enjoyed a promotion to President of the Reichsanstalt (the applied science institute near Berlin), in May. Lenard, promoted to become chief of Aryan Science, was ever proud of his fellow Nazi scientist. He wrote of Stark's promotion for a local paper. "It signifies a definite renunciation . . . of what might briefly be called Einsteinian thinking in physics."[70] In another venue, he bellowed, "We must recognize that it is unworthy of a German to be an intellectual follower of a Jew. Heil Hitler!"[71]

Lise Meitner, an established Jewish professor in Berlin, provided a common but now painful perspective on the upheaval of the time. "Here naturally everything and everyone is affected by the radical political changes," she wrote to Otto Hahn. "Last week already we had received instructions . . . to hoist the swastika flag." Of Jewish scientists, Meitner may have been the most influenced by Planck's inherent optimism. After first hearing the new Chancellor on the radio, she wrote that Hitler, "spoke very moderately, tactfully, and personally. Hopefully things will continue in this vein."[72] She maintained her post, working with Otto Hahn until 1938, but she always regretted her decision. "It is very clear to me today that I committed a great moral wrong by not leaving in 1933," she wrote after World War II, "since in the last analysis by staying I supported Hitlerism."[73]

When Hahn returned from spring 1933 travels, he was horrified. Most of his Jewish colleagues were less hopeful than Meitner, and many were making the difficult decision to leave Germany, despite uncertain prospects abroad. Although the traditional story has German Jewish scientists emigrating to the eventual Allies and helping to defeat Hitler, scientific jobs, like all others, were scarce in the depths of the Great Depression. Many scientific emigrants, if they ever did science again, became migrant hired hands for years to come, moving from lab to lab and nation to nation. The Law for the Restoration of Professional Civil Service, passed into law by Hitler on April 11, became the flashpoint for many. It prohibited Jews from government positions, except for those who had served Germany during the war. So while someone like Fritz Haber was exempt personally, he was ordered to purge his staff of Jews. Planck and Werner Heisenberg tried to convince their Jewish colleagues to simply wait out the passing rough period and to take advantage of the loopholes, but a wave of talent fled in 1933: most notably present or eventual Nobel laureates Haber, Max Born, and James Franck, a close friend of Lise Meitner's. Others left based on their conscience alone. Planck was particularly pained to see his heir apparent Erwin Schrödinger leave Berlin. Thirty years Planck's junior, Schrödinger was an outspoken critic of the Nazis, and he embarked on an awkward seven years of migrant work until establishing himself in Dublin.

A bleak sort of pragmatism guided Planck's actions during the Nazi years. A visiting Harald Bohr, brother of Niels, relayed that, "Planck was ... anxious to explain his reason of staying in his position ... was to try to do all in his power to help in the situation.... I understand that he is rather the only person, who may have some success in his efforts."[74] In fact, Planck had discreetly led worried delegations to the offices of Bernhard Rust (newly appointed as the Nazi's Prussian minister of cultural affairs), and Vice-Chancellor Franz von Papen, appealing for moderation.[75]

Planck then took the opportunity of his seventy-fifth birthday, April 23, to reply to Chancellor Hitler's friendly birthday card. He would request a meeting, talk about the future of German science, and put in a good word for his Jewish colleagues. A more quiet approach like this would be most effective, he was sure, and his last *public* foray into politics had been the most wrong-headed move of his life. Their private audience, on May 16, was devastating for Planck, but it gave him the truest measure of the new leader, and he no longer saw the troubles as a passing inconvenience.

Something more than pragmatism may have pushed Max in line, marching him open-eyed into a grim history. In 1918, Planck wrote to Einstein and wished openly for the Kaiser to step down and end the war. But he said he could never publicly request such an act. "I feel something that you admittedly will not be able to understand at all, . . . namely, a reverence for and an unshatterable solidarity with the State to which I belong, about which I am proud—and especially so in its misfortune—and which is embodied in the person of the monarch."[76]

So in May 1933, after his audience with Hitler, Planck did as he was told. He dismissed a quarter of the Kaiser Wilhelm Society's (KWG) senate, starting with non-Aryan members. And he then pledged his allegiance in meetings of the Prussian Academy of Science and the KWG. He said that no person could, "be allowed to stand, rifle at rest," as German science would rally to the new government.[77]

And although he noted for the KWG minutes that Einstein's work belonged with that of Kepler and Newton, with regrets he concluded that Einstein had only himself to blame for the permanent rupture between himself and the Academy.[78] Friendships, unlike dedication to the homeland, could be shed when duty called.

Shortly after Planck's encounter with the new chancellor, Otto Hahn tried to organize a protest on behalf of his Jewish colleagues, proposing a declaration with 30 signatures, but Planck refused. "If you bring together 30 such men today," Hahn recalled him saying. "Then tomorrow 150 will come to denounce them because they want to take their places."[79] Heisenberg also called on Planck, looking for guidance. They sat in Planck's living room, and Heisenberg recalled that Max's "smile seemed tortured, and he was looking terribly tired." Wangenheimstrasse 21 had been a temple of wisdom for younger scientists, with Planck their oracle, but he was changed. "You have come to get my advice on political questions, but I am afraid I can no longer advise you," Planck said. "I see no hope of stopping the catastrophe that is about to engulf all our universities, indeed our whole country. . . . You simply cannot stop a landslide once it has started."[80]

Fast-forwarding once more to 1944, Planck would see if his allegiance could be redeemed in the Nazi bureaucracy. After Hitler had barely escaped the assassination attempt, the frenzied Gestapo swarmed like ants over a damaged nest. Erwin was now in their illogical and all-powerful grip, amid a surge of political executions. What would Planck's years of cooperation be worth now, with the Reich cornered and the Fatherland on fire?

12

August 1944

Erwin's employer received a brief letter from the office of Heinrich Himmler, *Reichsführer* of the SS. The Otto Wolff Company had lobbied on Erwin's behalf, describing his importance not just to their operations, but also to international relations and the success of the Reich. They described him as the key person for their confidential international business (primarily obtaining metals for war machinery). But the reply from Himmler's office didn't stop at refusing their request. It cautioned them that further pursuit of Erwin's case, "would mean an undesirable burden" for the company. Shortly thereafter, the firm requested that Erwin resign his position, and he complied from his jail cell. This left Nelly with no income and an actual debt, as his former employer claimed they'd overpaid Erwin in royalties.[1]

Max Planck decided to travel to Berlin, despite the condition of the city, to join forces with Nelly—or *Nellchen*, little Nelly, as he called her now. In a letter to Max von Laue, he said he had trouble wrapping his head around the case. Erwin must have been arrested just because he knew some of the would-be assassins—that's all. And Planck said he took comfort in the fact that so many people were arrested, including some just as prominent as Erwin.[2]

Max and Nelly decided to triangulate on Heinrich Himmler in part because of family connections. Nelly's sister knew Himmler's wife and wrote two letters beseeching Frau Himmler's influence with her husband. Meanwhile, Max wrote to Herr Himmler directly.

> I was informed by my daughter in law that my son Erwin was arrested 23 July and that his situation is said to be very serious. Because of the deep relationship with my son, I am convinced that he had nothing to do with the events of 20 July. I am 87 years old and am relying in every regard on the help of my son. Until today, I have strived to live my science and my honorary positions

> and, in this way, to serve my Fatherland. I was only able to do this because of my son's assistance in all affairs. At the end of my life, this son is the only one remaining from my first marriage.

To underline Erwin's importance, he makes a distasteful addition, hopefully without Marga's knowledge.

> My son from my second marriage is not intellectually fit to carry on the family standards, while Erwin's character and his gifts embody everything from generations of our family. I am asking you, honorable Herr Reichsführer, to put yourself in my situation and try to understand what it would mean to me, while also considering the prestige of my name in Germany and abroad, if I had to lose my son due to a harsh sentence.[3]

By mid-September, they had their answers. A letter from Himmler's staff informed Max that the charges and evidence against Erwin were so strong "that a release is impossible." Meanwhile, the Reich still denied Max and Nelly's petition to visit him. Their only small comfort came in the form of Erwin's short letters from prison, sending his love, saying he was fine, thanking them for the good thoughts, but not saying much about his imprisonment. He requested a few items that would make his stay more bearable, like a copy of Homer's *Odyssey*.[4]

In truth, Max had no good options for sympathy in the Reich. Despite pledging his support for the Nazis in 1933, he never joined the party, and over the next decade, he suffered increasing scrutiny and criticism from Reich stalwarts.

His summer 1933 report for the Kaiser Wilhelm Society (KWG) shows him playing by the new rules. The Reich requested he dismiss twenty-seven Jewish assistants of the society. His reply broke these into three groups: nineteen he dismissed (yes, as you wish); three questionable cases (maybe, but this is complicated); and five "cases of hardship" (best if we keep them, for the good of the society). But by the end of July, Planck told several directors he had failed to obtain some of the exemptions.[5] Records show him continuing to campaign for certain people, predominantly those staff within the family trees of well-known scientists. He tried to obtain official exceptions to the new laws or even Aryanize individuals like Mathilde Hertz, a relative of the great Heinrich Hertz. His efforts for Mathilde lasted two progress-free years, and she left for England in 1935.[6]

Meanwhile, Planck continued to pitch the Nazi nastiness as transient, like some physical effect that would fall exponentially with time—"a thunderstorm which would soon pass away," he liked to say.[7] One Jewish scientist heeding his advice was the biochemist Carl Neuberg. While chemist Walther Nernst advised Neuberg and others to leave in 1933, many followed Planck's advice instead. "He said I was protected as an old civil servant of the Kaiserreich," Neuberg wrote. "Later I was forbidden to leave the country."[8] Neuberg eventually escaped, and though he survived the Nazi era, he looked back on Planck's advice with bitterness.

In the following year, Planck began to push back using a narrative of pragmatism. He noted that extreme acts and policies within Germany would hurt the Fatherland's international reputation. He particularly spoke out against any government funding for eugenics and related pseudoscience that would reinforce growing disdain and disbelief in other international quarters. Such reactions abroad had real consequences for Planck. He'd been working with the Rockefeller Foundation to finally design and build the Kaiser Wilhelm Institute of Physics. The Institute's meeting minutes of the time underline his worries: "Privy Councilor Planck and the undersigned stated that it was necessary that . . . nothing happen in German science that might disturb the Americans, because undoubtedly a certain reticence can be felt at this time."[9] In the same meeting, Planck emphasized the emerging importance of atomic research, following the revelations of Hahn, Meitner (in exile), and others.

We glimpse a biased but telling view of Planck in 1934 from the diary of Lotte Warburg, daughter of the deceased physicist Emil Warburg. "Why does he stay on and allow himself to discharge people from the KWG? . . . Why does he go about hunched over, moaning and complaining, instead of throwing back his head and damning them all?"[10]

He wasn't shouting down the Reich, but he wasn't curling into a ball either. In January of that year, he organized a memorial against government orders. The Jewish chemist Fritz Haber had died in exile, and although the Nazis forbade any civil servant from recognizing Haber, Planck gathered von Laue, Otto Hahn, Lise Meitner, and several others to pay their respects. "Haber kept faith with us," Planck said. "We shall keep faith with him."[11] Later that year, Planck again refused to supplicate. When President von Hindenburg died in August, Hitler quickly moved to merge the roles of the president and chancellor, cementing a dictatorship. Johannes Stark drafted a proclamation of support, but Planck and von Laue refused to sign.[12]

In 1935, Max continued to sharpen his arguments about the utility of science for the Reich. His new KWG reports began emphasizing the role that science could play in international relations, removing rough edges and negative impressions. "Many a misleading view abroad can be rectified," he suggested.[13]

At the same time, he clearly realized that negative views were not misleading at all. In 1935, physicist James Franck invited Planck to Denmark. "No, I cannot travel abroad," he replied. "On my previous travels I felt myself to be a representative of German science and was proud of it. Now I would have to hide my face in shame."[14] He launched a new public lecture, "Physics and the Struggle for a World View." Lise Meitner later remembered this one with admiration, as he underlined that "justness is inseparable from devotion to the truth." Planck linked the enterprise of science to the underpinnings of a healthy society. He warned audiences against becoming a community in which, "the guarantee of justice begins to vacillate," and against a system in which rank or heritage affect legal protections—this was not a subtle jab.[15]

And so from the earliest time of the Nazis, we see a consistent approach from Max Planck (Figure 12.1). While he would never openly challenge the government, he never fully aligned either; he continued to make statements between the lines and to defy them in quiet ways. He saluted the Nazi flag and signed official letters "Heil Hitler!" but he also preached international cooperation and pursued amnesty for his Jewish colleagues.

In early 1936, he feared losing his balance on this razor's edge when the American press threw a spotlight on his sentiments. In an article titled "The Last Stand," the *New York Times* reviewed one of his public speeches and painted him as a bold intellectual, standing against the Reich from within. "Max Planck, to his everlasting glory, went as far as common sense permitted." They translated his public message to be that, "personalities and brains count for more than race or totalitarianism." Planck was horrified, fearing what a backlash might mean for his Jewish colleagues at the KWG.[16]

One way or another, the Reich noticed Planck as a public persona resisting their *Gleichschaltung* movement. The word literally means "equal switching," and was used by the Nazis to suggest the mass alignment of all German citizens to the same goals, swearing allegiance to their country and party (increasingly synonymous). In March 1936, for instance, two enormous new zeppelins, the Graf and the Hindenburg, loomed roaring over the country dropping leaflets with voting instructions. The upcoming referendum would place Nazis throughout the Reichstag and support

Figure 12.1. Max Planck delivers a talk in Stuttgart on June 24, 1935.
Courtesy Archiv der Max-Planck-Gesellschaft, Berlin-Dahlem.

Hitler's alarming international bravado. Any opposition had to be silenced, even minor irritants like Planck.

Ernst Gehrcke, the physicist who had joined Weyland's anti-Einstein stage show in 1920, emerged now to revisit Planck's famous discovery. He published an April 1936 article entitled "How the Energy Distribution of Black-Body Radiation Was Really Found," in *Physikalische Zeitschrift*. Much more dangerous and credentialed than the Weylands of the world, Gehrcke made for an insidious adversary. He wrote a detailed alternate history in which the experimenters of 1900 had actually mapped the perfect formula and then shared it with Planck. "It is characteristic of the physics of the past epoch that Planck's elementary mathematical trimmings were esteemed more highly than the original, fundamental physics discovery by Lummer and Pringsheim."[17] For Nazi sympathizers, that epoch meant the deluded "Jewish Physics" dark ages. According to Gehrcke, Planck was only famous because he had supported the public relations machine of Jewish scientists. Meanwhile, Stark began to repeatedly attack Planck in public, calling him a "nefarious influence," and one of Einstein's "creatures." Planck, "showed himself to be so ignorant about race that he took Einstein to be a real German who should be honored in his country." Lenard joined the chorus, calling the entire KWG a "Jewish monstrosity." With Planck as president, Lenard said the society should be disbanded.[18]

The Reich spoke an increasingly negative line against science. In the fall of 1936, Hitler instituted a new law forbidding any German from accepting a Nobel Prize. The value of such awards had plummeted in the Führer's eyes when the German pacifist and concentration camp prisoner Carl von Ossietzky won the Nobel Peace Prize that year.[19]

Planck clung to his post, hoping to still do some internal good, and the attacks continued. In 1937, the SS newspaper *Das Schwarze Korps* labeled Planck a "white Jew" for following in the footsteps of Einstein. And now he would be investigated for rumors of Jewish heritage—the charges moved from Max being a "white Jew" to, perhaps, a one-sixteenth actual Jew.[20]

Meanwhile, the drumbeat of war grew louder. Agricultural lands were repurposed to the military, and food rationing began. Max von Laue surveyed the landscape of his homeland and found a way to send his only son to America.[21] (Theodore von Laue would go on to earn a PhD in history at Princeton and make pointed criticism of his father and other German scientists after the war.)[22]

Just a year from the outbreak of World War II, Planck presented the 1938 Planck medal to the young French physicist Louis de Broglie, a double insult to Nazis, as he was not only French but also following the ways of "Jewish Physics." With a presentation to the French ambassador, Planck said, "May a kind fate grant that France and Germany come together before it is too late for Europe." The ambassador replied with words that echo in Planck's long legacy. "In Herr Geheimrat Planck we honor one of those accomplished men of whom not only his country but the entire world has a right to be proud."[23]

Planck, now 79, decided to resign as president of the KWS, but he maintained his post as secretary of the Prussian Academy of Sciences. (Others of the old guard who had never joined the party also stepped down. His neighbor Karl Bonhöffer, for instance, retired from the Department of Psychiatry and Neurology. An SS officer took over and started euthanizing psychiatric patients.) That fall, the University of Berlin banned Jewish students from attending the regular physics colloquia, and Planck was asked to dismiss three Jewish colleagues from the Academy. He did as he was asked and wrote to von Laue his relief that the three, like the Jewish students, hadn't made a spectacle of the awkward situation.[24] By the end of 1938, Meitner had fled, the Nazis owned the Academy, and Max had resigned his last official position. At 80, he was finally and completely retired, having set each of his many official hats aside.

His eightieth birthday provided a bright spot of levity. His friends and colleagues came together for a "great celebration of the physicists" hosted at Berlin's Harnack House on April 23. The party featured a one-act farce about the history of quantum theory. Physicists including Peter Debye, Arnold Sommerfeld, and Werner Heisenberg prowled the stage.[25] Also delighting Planck that year was the opening of the new physical home of the Kaiser Wilhelm Institute for Physics. Its large, windowless tower housed new particle physics experiments. Electrified to one million volts, instruments running the length of the tower could observe the creation of electron-positron pairs.[26] Institute director Peter Debye christened this the Planck tower, and today, it houses archives of the Max Planck Society (formerly the KWG) in southwest Berlin.

But the world outside the temples and towers of science was less appreciative of Planck. Despite his stepping away from work and spending more time with Marga, Nelly and Erwin in the mountains, the public attacks against Planck continued, but now more from the likes of the Reich Student Command, a Nazi youth organization. Their newspaper labeled Planck "founder and ringleader" of the nation's wrongful turn to mathematical physics, "which rules out any real thinking."[27]

Planck saw the march to another war as a "sadly grave" time for his Fatherland. "I, for one, see in current developments basically only a senseless self-decapitation of the Aryan-Germanic race," he wrote to a colleague in 1940. "Only the gods know how this will end."[28]

From our great remove, one can certainly ask why Planck stayed or, having decided to stay, why he failed to do more given the evidence around him. Why would he even agree to lecture in occupied territories, no matter the content, where in plain sight German forces exerted such a cruel and obvious grip? A typical reply would underline his pragmatism and his age. He was nearing the end of his career and the end of his life, thinking he could do more good quietly inside than he could do loudly outside. (As Carl von Ossietzky allegedly said when opting to stay in Nazi Germany, it is a hollow voice that speaks from across the border.)[29] But setting age and pragmatism aside, two stories leaven an emerging picture.

In 1907, when Max Planck swam in familial happiness and the giddy early days of Einstein's relativity, he sat with an enticing job offer. After Ludwig Boltzmann's suicide, Planck interviewed in Vienna as a possible replacement, and it was during this visit that the young Lise Meitner heard him give a lecture. His interview went well enough that the University of Vienna extended an offer, and he faced a difficult decision. At age 49, had he accomplished all he could in Berlin? Wouldn't it be wonderful to be closer to the beloved Alps? And immersed in Vienna's music? Would a move benefit or harm Marie's fragile health? And what were they going to do with their increasingly empty Grunewald nest anyway?

But Planck opted to stay. He identified too strongly with the empire that had grown with him to maturity, success, and esteem. And all signs suggested the best was yet to come. If he won the Nobel Prize, as many suggested, shouldn't he receive that in Berlin? The news of his decision spread quickly in Berlin, and in the middle of a dark Grunewald night, Marie and Max woke to singing outside their home. When Max went blinking to the door, he heard a song from his youth, a sailing song of *Kielers*, coming full throat from a mob of torch-bearing students. Beer breath, spittle, and testosterone must have hit Max Planck in gusts. The students came to show their love for their professor and cheer the good news. And when they finished their song, he confirmed he was staying. He shouted into that feral night that he would never leave.[30]

The second story illuminating Planck's Nazi-era behavior doesn't involve him at all. The Swedish journalist Gunnar Pihl relays a common story from the middle and late 1930s. It exemplifies the kind of tale Germans traded in hushed tones during the years of the Third Reich, and certainly, Max Planck had heard it or stories much like it. As Hitler consolidated power and held sham votes along the way, one Catholic clergyman told his congregation, "I, Bishop of Münster, have voted against Hitler." He said he must stand against obvious evil, and he was arrested later that day while the pews were still warm. But the town of Münster was the boyhood home to German hero and flying ace Colonel Werner Mölders. When Mölders heard of his bishop's arrest, he gathered his party membership card with all his war medals and mailed them to a Nazi party office. Facing a rebuke from the popular hero, the Nazis quickly released the bishop and returned the materials to Colonel Mölders. So a prominent figure could and did stand up to the Nazis. And within a few weeks, the colonel was dead, allegedly killed in a routine solo flight. But those near the airfield reported no wreck, only the horrific sight of a body falling to earth, pushed from a plane far overhead.[31]

Planck was caught between an incredible allegiance to and identification with Germany, on the one hand, and the practical reality of what could be accomplished in protest, on the other. This neither excuses his actions nor completes a measure of his conscience, but by carefully mapping his boundaries, we take another step toward understanding what he did and didn't do.

Planck had little direct contact with the Reich's machinery from the time of his resignations until the time of Erwin's arrest, roughly six years later. He threw himself into his philosophical lectures, his correspondence, and the remnants of his family, including Erwin, Nelly, and granddaughters Grete Marie and Emmerle. In 1944, he was ill equipped to gauge, let alone affect, the internal dynamics of the Third Reich.

Perhaps he and Nelly focused their efforts on the family of Himmler because they had heard of other cases where familial sentimentality thawed his ruthless Nazi gaze. Heisenberg, for instance, found himself under public attack in 1937—he was yet another deluded Einstein worshiper according to Stark and the SS. Heisenberg's mother wrote a note to Himmler's mother, as if the entire episode had been a misunderstanding on a childhood swing set, and the letter worked magic. Himmler wrote to Heisenberg directly, admitting, "You were recommended by my family." He assured the physicist, "I have put a stop to any further attack on you."[32] Heisenberg's reputation was not assailed further (at least in Germany), and he would go on to lead the German uranium project.

But Nelly and Max had no success via Himmler's wife. The Planck clan tried to maintain one another's spirits—Nelly sugar-coated things for Max where she could, and, as in the following letter, he in turn chided her for working too hard for the war effort, including running an early version of an X-ray machine for the wounded.

> I knew from the beginning that your work is exhausting and also dangerous because of the radiation. But I think it is irresponsible that you additionally put your health at risk with your constant blood donations. You are usually so smart and prepared for all eventualities, but in this case you are like a child who cannot consider the future. What do you gain from all your sacrifice, when Mops will return to find, not his normal chubby wife, but a skeleton, somehow alive but without her normal charms? . . . please, dearest little Nelly, take this advice to your heart from your father who is worried about both of you to the same extent, and look forward to the moment you embrace your Mops again.[33]

Even as Planck made requests of the Reich, he made little effort to endear himself to the Nazis. He missed a mid-October deadline to submit a requested propaganda testimonial, one for the "Confession to the Führer" series. He finally just wrote, "I am sorry to tell you that because of the imprisonment of my son, I am unable to find the words that would be consistent with the purpose of this brochure."[34]

On October 23, 1944, Erwin went to trial, or what passed as such. His official charges declared:

- Preparing high treason,
- To alter the Reich with violence,
- To attempt to remove the constitutionally provided powers of the Führer,
- To establish a group preparing high treason, and
- Abetting hostile forces during a time of war.[35]

The "People's Court" hosted the case. Defendants here were often made to wear court-provided clothing: a shabby jacket, no tie, and for some, oversized pants with no belt. Judicial president Roland Freisler ruled here, shouting down defendants, asking rhetorical and humiliating questions. The word "shouting" cannot convey the absurd violence of Freisler's voice. The court in action is readily and sickeningly available for viewing today, as the sessions were recorded both for propaganda and Hitler's viewing pleasure.

If Erwin looked about the grim room, he saw only party officials and a few Nazi propaganda "journalists." Family and friends of any sort were forbidden from the chambers. In the brief "trial," Erwin and his lawyer built a narrative of Erwin always trying to be a good German. He had told the conspirators he just wanted to avoid the horror of 1918. He worried about what would happen when the Allies overran them. When others approached him discussing a diplomatic path to peace, he refused them, but he thought it his patriotic duty to listen.[36] Such an argument won no sympathy from Freisler, who pronounced Erwin's death sentence and moved to the next case.

Two days after the verdict, Max Planck sent a letter directly to Hitler.

> I am deeply shaken by the news that my son Erwin was sentenced to death by the People's Court.
>
> As you've repeatedly commended me—in the most honorable expression of recognition—for my achievements in the service of our Fatherland, I feel entitled to ask that you lend a sympathetic ear to an 87-year-old.
>
> As thanks from the German people for my life's work—which has become an everlasting intellectual asset for Germany—I ask for my son's life.[37]

If Planck received any response, it has not survived. As von Laue wrote to Lise Meitner in Stockholm, "one has not much hope."[38]

Planck also wrote directly to Himmler once more. He attached his earlier letter and now pleaded for the commuting of Erwin's sentence to life in prison. Erwin's company risked irking the Reich and wrote a series of new letters on Erwin's behalf. Nelly, along with her friends and family wrote to an array of Nazi officials, including Franz von Papen, the former Weimar-era chancellor who had somehow appeased his new overlords. In this case Nelly's letter, co-signed by Max Planck, received a personal reply.

> Madam!
>
> I received your letter today and was very shocked by the news of your husband's sentence. I was convinced that he completely withdrew from all political activity after that business in 1934. Due to the direct orders of the Führer, I am absolutely prohibited from filing or supporting clemency pleas—so I regret that there is no way I can assist. But even though I am completely ignorant about this case, I might hope—due to the great standing of your father in law—that the Führer might grant your clemency plea. I hope you will be spared the most difficult outcome.[39]

13
November 1944

Planck's lifelong motto was *Man muss Optimist sein*, or "one must be an optimist," but in late 1944, he must have let himself consider the unthinkable. He and Nelly had stretched and then exhausted their contact lists in the Reich. Only two questions remained. Would a person in power suffer a whim of sentiment? Failing that, would a Nazi underling risk his own standing to speak for the son of Max Planck? Max would have looked on the odds of Erwin's release like Ludwig Boltzmann had once looked at those rare cases where a molecule bounced left instead of right and, for a split second, entropy in the universe took a tiny dip instead of marching ever upward—an event practically but not statistically impossible. And even if Erwin against odds evaded his sentence, the odds of surviving the war looked worse than ever. Erwin had already escaped an Allied bombing; it was, "really like a miracle," Max wrote in a letter. Explosions obliterated one building, while next door, Erwin's prison wing stood untouched.[1]

Max wrote to Nelly on the first day of November.

> Dear Nell! Never in my life have you been so close to me and have I felt so tightly connected to you as in these days, when both of us turn all thoughts and worries towards the one, who, of all the world's men, is closest to us, who gives us the greatest security and foothold in the world, and who is to be taken away from us. . . . Let's continue hoping that fate smiles on him again. Hopefully he stays healthy while imprisoned and can sleep properly, even without alcohol, his usual sleep tonic.[2]

In the closing months of 1944, streams of Berlin refugees swelled Rogätz to 10 times its prewar population. Max was still regularly seen walking through them on his way to a daily shave. Moving among these throngs, he must have occasionally seen a kindly pug dog face—*Mops*! In this Erwin vigil, subconscious excitement would surrender to conscious dread. And if some of the refugees recognized the celebrated scientist? His stooped and painful gait mirrored the empire's expiring dreams, hatched when the fearless Max Planck stood erect, and demanded answers from the universe.

But in 1944, even his trusted dialogue with nature—halting, sometimes confusing, but always following certain rules, and always progressing—lay unrecognizable, like so many smoldering cities.

The key moment illustrating physics passing Planck took place in 1926. At the age of 68, he at first welcomed a new take on quantum theory called "quantum mechanics," dreamed up by a young group of physicists from Göttingen. He was proud that the emergency funding apparatus, the *Notgemeinschaft*, had kept these young guns blazing. But in that year, he publicly announced that, whatever came next from the new theory, physicists could rest assured that the bedrock foundation of physics stood firm. Namely, they could still rely on experiments that took a true measure of natural world, without affecting it. That is, the exact wording of their laboratory questions would not affect the answers nature uttered in reply.

Physics had long relied on the notion that particles were being observed discretely, like so many animals in an enormous nature preserve. With experiments functioning as silent telephoto lenses, we could count on observing a particle's natural state. But would the new brand of quantum theory render particles more like animals in a zoo, with behavior inseparable from the confines of the experiment? Could humans really trust what they observed as fundamental reality or were they looking at byproducts of their intrusions? In 1926, Planck reassured everyone that physics rules out that worry, "from the very beginning."[3] He couldn't have been more wrong, and his timing couldn't have been worse.

Within months, physicists started a new dialogue with the microscopic world, a dialogue marked by a bizarre mix of unprecedented predictability, on the one hand, and a chilly intellectual distance, on the other. Not only did the new physics admit that experiments could influence their subjects, but it also drew a curtain across the bars of the particle cages. It was as if the young physicists said, "Nature, we will make you a deal. As long as you act predictably, we will quit asking *why* you do things," and nature readily accepted. The treaty persists in large part to this day, despite the protests of Planck, Einstein, and others.

From the time of his first scientific work onward, Planck had been fixated on nature's motion from starting points to endpoints. In his 1879 doctoral thesis, he reworded Clausius's second law of thermodynamics to implicitly include time's arrow: *The process of heat conduction cannot be completely reversed by any means.* Nature somehow prefers the end state of a process, Z, to the starting point of the process, A. The cold cup of coffee never spontaneously regains its warmth as it sits forgotten on the counter. To Planck, thermal physics provided a windsock by which humanity could see the warm breeze of time, always blowing in one direction. Planck never lost his interest in "irreversibility," and he believed that attempting to fully understand it was among the most important tasks for physics. In fact, he had put his finger on a concept that still plagues us nearly 150 years later. How does one reconcile beautiful and successful physical theories (like Newton's laws of motion or Maxwell's equations for electromagnetic waves), which show no preference for a direction of time, and somehow then derive a more comprehensive theory in which time *does* have a preference, as per reality? Most theories in physics would work just fine if you put them in reverse. Just like every up has a down, and every left a right, every positive axis of time has a negative axis trailing behind it, mathematically enticing real estate with no bids from the real world.

Planck longed to find some cause—a hidden archer launching time's arrow. In 1897, he labeled this "the fundamental task of theoretical physics," and it drew him further into black-body radiation. He became convinced that, in this unique problem, he would prove an irreversible pattern using only established and nonstatistical approaches. In that year, he published a five-paper series on the black-body problem called, "On Irreversible Radiation Processes," but he couldn't prove his title's seductive notion.[4]

Instead of divining why time moves in one direction, he inadvertently opened a new field, quantum theory, with its own questions that would consume the attention of physicists for decades to come. He asked nature why time moves in one direction, but nature replied that he would be better off worrying about matter and energy.

Planck expected only glacial progress on these new questions. Einstein aside, few others made advances in quantum theory for the first decade after Planck's discovery. In 1912, he concluded the preamble to his second quantum theory with a prediction of more slow going. "Any one who, at present, devotes his efforts to the hypothesis of quanta, must, for the time being, be content with the knowledge that the fruits of the labor spent will probably be gathered by a future generation."[5]

But before this book hit the shelves in 1913, a Danish graduate student sidestepped Planck's morose prediction (Figure 13.1). By March of that year,

Figure 13.1. Niels Bohr circa 1911.
Courtesy the Niels Bohr Archive, Copenhagen, and courtesy AIP Emilio Segre Visual Archives.

Niels Bohr presented a wholly new vision of the atom. He used the notion of discrete energies to shackle electrons within an atom. Now, instead of having any value of energy, the electrons had only a set of allowed *quantized* options—one, two, or three chunks of energy, but never anything in between. The model immediately provided, for instance, a conceptual way to understand the colors of fireworks—why different elements emit specific colors when excited or ignited. To be sure, Bohr's model was not perfect, and fine-tuning would follow, but physicists began to realize a new era for light, energy, and fundamental matter. The number of publications involving the quantum idea nearly doubled from 1912 to 1913.[6] The teenaged theory wasn't just putting up a few posters—it wanted to knock out some walls.

Fresh from his atomic model success, Bohr in 1917 began to build what would come to represent an impregnable fortress for quantum theory. His Institute for Theoretical Physics opened in Copenhagen four years later.

There he regularly hosted the best and brightest theoretical physicists from around Europe and America for stints of intensive work and collaboration. Over the next decade, visitors included Werner Heisenberg, Max Born, Paul Dirac, Wolfgang Pauli, and Erwin Schrödinger, among others—the eventual prophets of modern quantum theory.

Bohr won the 1922 Nobel Prize in Physics for his brilliant 1913 "Trilogy" of papers concerning a quantum atom. The same year also settled a long-simmering question for Planck and most other physicists. Einstein had first suggested his *lichtquant*, or light quanta, in the miracle year of 1905. Planck rejected this notion outright and even had sincere doubts as late as 1918. As he wrote to his friend Hendrik Lorentz that spring, "the question whether light rays themselves are quantized or whether the quantum effect occurs only in matter is the first and most serious dilemma confronting the entire quantum theory."[7] In fact, very few physicists had accepted the reality of such light packets, but measurements from the American laboratory of Arthur Compton confirmed their existence by the end of 1922.[8] (Even then, neither Einstein nor Compton had ever used the word "photon," which appeared for the first time in a 1926 paper by the chemist Gilbert Lewis.)[9]

So now physics faced an uncomfortable fact that Einstein had suggested for some time: Light was composed of particles, but the particles could exhibit the character of waves. "Wavicles" are not much easier to grasp for the human mind now than they were in the early twentieth century, but it may be best understood as a vocabulary problem. The reality of the photon, unlike any object or quality of our everyday lives, is analogous to describing an artichoke to a person who's never seen one. For size, weight, and color, you might describe it as "kind of like a pear," but in terms of skin and texture you might say it's "kind of like a pineapple." The photon is most certainly an artichoke, but we humans are stuck trying to understand it as either pear or pineapple, so we force inappropriate concepts and constraints on nature. For the nano world, the images of billiard balls or ripples on a pond are not quite right, but these are the clumsy tools we carry from the macro world.

Work in the field of quantum theory shifted into higher gear in the 1920s. A French doctoral student Louis de Broglie wondered, since light waves could have the properties of particles, couldn't particles also then act like waves? In essence, he asked the world of physics: What if *all* these tiny, invisible things are artichokes? Not knowing what to think of such a radical idea, his thesis advisor sent a copy to Einstein, and Einstein blessed it immediately.[10]

Meanwhile, Werner Heisenberg was leading a group to introduce a new quantum theory to the world. In 1925, he fled his Göttingen hay fever and retreated to Helgoland, a barren rocky island in the North Sea.[11] Just as 23-year-old Isaac Newton had fled a London plague and made critical advances amid bucolic scenery, 23-year-old Heisenberg penned the start of quantum "mechanics" in stark surroundings. Newton's classical mechanics had described macroscopic objects like billiard balls, whereas quantum mechanics would do the same for something like de Broglie's wavy electrons. Heisenberg's version is still taught as "matrix mechanics," since an electron's properties, like energy, were for the first time partitioned over the face of a tic-tac-toe–like matrix of options. At least mathematically, an electron could now retain many options simultaneously. Some of the electron's reality was over here, and another portion was over there.

At the same time, an alternate version of quantum mechanics arose from the mind of the young Austrian physicist Erwin Schrödinger. He published his "wave mechanics," in 1926, a year before assuming Planck's position at the University of Berlin. Planck followed Schrödinger's work with great interest, writing letters about next steps to both Schrödinger and Hendrik Lorentz.[12] Planck hoped that Schrödinger's theory offered a *deterministic* future to quantum mechanics—meaning A would still lead to Z somehow, instead of having just some probability of Z, mixed with a little probability of Y, and so on. Planck desperately wanted a quantum mechanics that could describe the path of an electron with the same confidence that a center fielder traces a flying baseball. Instead, some of the new quantum ideas threw many baseballs at once. All realities were valid and not even the baseball players could tell them apart until the moment the ball appeared in just one glove, with all other gloves, raised aloft, coming up empty.

Although Planck may have seen the writing on the wall, or in the matrices, he doubled down on his lifelong fundamental beliefs about humans, science, and the natural world. In the two aforementioned 1926 lectures, he assured audiences that, no matter where quantum theory went next, science would not have to worry about an experiment influencing its own subject, meaning that scientists would still be free to observe nature as if they were sitting in a perfectly camouflaged and scent-proof hunting blind—nature would behave as if nobody was watching. In the same year, he wrote a series of six letters to the philosopher Theodor Haering, explaining the importance of his stance. Planck claimed that picturing a universe where data are collected without altering the system under study, "is the basic presupposition of any sort of scientific knowledge."[13] Indeed, Planck must have at times viewed Heisenberg and his gang as barbarians who threatened to ransack science back to darker times.

In late 1926, Max Planck was 68 years old, while the average age of the new quantum innovators was just under 28. (I average a defensible but definitely debatable list of Bohr, Heisenberg, Schrödinger, Max Born, Paul Dirac, Pascal Jordan, and Wolfgang Pauli.) Youths were running with Planck's ideas and taking them well beyond his comfort.

Einstein, Planck's ally, watched the developments with great skepticism. The still creative and convention-free genius would never be an "establishment" physicist clicking his tongue at new ideas. But he also had to recognize the natural rise of a new physics generation. Shortly before turning 40, Einstein lamented that one naturally becomes "more *blockheaded*" with age and experience. Although many of us feel the slowing or perhaps stiffening of our minds, few have expressed it with Einstein's playful eloquence. "The intellect gets crippled," he wrote in 1918, with the ink barely dry on his general theory of relativity. "But glittering renown is still draped around the calcified shell."[14]

Einstein initially sought to dismiss Heisenberg's matrix mechanics, writing to Ehrenfest that Heisenberg had "laid a big quantum egg." Even though the Göttingen crowd may believe their strange work, he wrote, "I don't."[15] He locked his jaw and sharpened his pencil for battle.

By the end of 1926, three parallel versions of quantum mechanics had emerged. To Schrödinger's wave mechanics and Göttingen's matrix formulation, an English graduate student named Paul Dirac added a third and arguably most elegant version in his PhD thesis. The next tectonic shift in quantum theory arrived in 1927, when Heisenberg published his famous uncertainty principle, giving an electron's position and speed a new sort of relationship. Given a wave–particle duality and thereby a more hazy or furry vision for an electron, one had to accept uncertainty in both its position and its speed—meaning the electron would inhabit a *range* of position and speed values instead of one exact value for each. Instead of being able to put one's finger on the electron's location—it's *right here*—one would have to use both hands held apart—it's *somewhere in here*. Moreover, the respective uncertainties of position and speed were now linked together. If a clever experiment narrowed the uncertainty in an electron's position, the same experiment necessarily increased the uncertainty in its speed. When a scientist brings her hands together to locate the electron within, it squirts out like a watermelon seed with some new and unforeseen speed. The very probing of the electron, using light or anything else, would affect its state of being.

Shortly after reading Heisenberg's paper, Planck wrote to Lorentz, and the two aging physicists could only shake their heads at what was happening. Max called the "ominous" uncertainty principle "an unacceptable limitation of the freedom of thought, and . . . a mutilation" of the sacred notion of pure and simple measurements.[16] But in the end, Heisenberg helped the world better understand the true nature of Planck's own constant.

When Planck had speculated in 1908 that spacetime was quantized, he was on the right track,[17] and with our long hindsight, Heisenberg's 1927 work shed the brightest light on Planck's "quantum of action," *h*. The uncertainty principle sets a limit to the possible precision of physical knowledge for a particle or a process. If we multiply the uncertainty of a particle's position by the uncertainty in its momentum (or alternately the uncertainties of energy and elapsed time for a process), the result, as ordained by the universe, can *never* be smaller than Planck's constant.[18] In one sense then, *h* should be viewed as the fundamental grain size to the *knowability* of the physical universe. Planck himself eventually said that *h* "erects an objective barrier" between our experiments and ultimate precision, such that "progress will only give this barrier even sharper outlines than it had before."[19]

Following the uncertainty paper, Heisenberg, Bohr, and Max Born fell into grueling round-the-clock discussions as to the nature of the emerging quantum reality. Allegedly, in their first meetings, Heisenberg was reduced to tears.[20] They eventually found enough common ground for a truce of sorts. That fall, the three men set out a new framework for what came to be called the "Copenhagen interpretation." First, Bohr presented his principle of "complementarity," meant to draw a bright line between the musings of the human mind on one side, and the complexities of the atomic world on the other. The pristine "natural" state of something like an electron, according to Bohr, included two equal sides like a coin (a wave side and a particle side), but the inner truth, balanced on the coin's edge, could never be captured by the ungainly hands of human science, which typically bumped an electron enough to make it look mostly like a particle or mostly like a wave.

Heisenberg and Born, claiming they had *completed* quantum mechanics, unveiled their theory's statistical underbelly, employing the Heisenberg uncertainty rule and declaring that the "waves" of Schrödinger's theory were in fact just mathematical ripples of probability—the peak of a wave just showed the point of highest probability for a particle's existence. With Schrödinger's approach fully assimilated, they snuffed Planck's hope for a more comfortable quantum future.

The Copenhagen interpretation has tempted many observers to run with it and claim scientists can no longer know or predict anything whatsoever, but nothing could be more incorrect. In terms of its experimental predictions, quantum mechanics and its descendants are still unsurpassed.

Quantum theory's most compelling glory was its description of an element's "atomic spectrum," the specific set of light colors that it gives off under duress. The aforementioned case of a firework gives an example, in which different sizzling components give us different colors; a neon sign provides another. This is very different from the universal, atom-*independent* thermal radiation that Planck described. The distinct atomic spectra are essentially the fingerprints of underlying atomic architecture. The spectra shine forth only when an element is agitated by the outside world, whereas thermal radiation is ever-present, based only on temperature. Before quantum theory, the spectra were complete head scratchers. Quantum theory, starting with Bohr's atom and moving through decades of improvements, learned to predict and understand these exact colors with ever-better and eventually ridiculous precision. At some point, even skeptics had to nod, sigh, and say, *okay, this must be right.* No theoretical model has been as precisely and eerily successful in describing nature's behavior as quantum mechanics and its descendants. At the same time, perhaps no other model has admitted its own intellectual limitations as bluntly: The authors of Copenhagen drew a bright line between describing the *measurements* of evidence like the spectra, and actually claiming to *know* what exactly was happening under the atomic car hoods.

The Copenhagen interpretation threw up its hands at a complete blow-by-blow understanding of the subatomic world. It even cordoned off that terrain from curious human beings. Bohr said bluntly, "It is wrong to think that the task of physics is to find out how nature *is*." Rather, he claimed the task of physics was only to find what humans could reliably *say* about nature.[21] Well into the twenty-first century, many physicists still consider the Copenhagen interpretation our best and most honest assessment, whereas others are itching to challenge it. Physicist and Nobel laureate Steven Weinberg has arguably spent as much time thinking about quantum theory as any other living person. In his recent textbook, *Lectures on Quantum Mechanics*, Weinberg discusses how we can interpret the *meaning* of the wonderfully useful and successful machinery of the theory. He admits that, "it is hard to live with" a framework for computing predictions without a deeper understanding of these microscopic systems. "My own conclusion (not universally shared) is that today there is no interpretation of quantum mechanics that does not have serious flaws."[22] We may yet get to an underlying and more satisfying truth.

Planck's friend and one-time assistant Max von Laue shook his head at the Copenhagen interpretation, with its focus on being finished and washing its hands. "Planck has mentioned conscientiousness and loyalty as the necessary character traits of a scientist," he wrote. "I think that we should add patience." And he would always see the Copenhagen interpretation as tinged with Europe's pessimism after the Great War.[23] It was as if younger minds had decided that humans, judging by their senseless brutality, were not smart enough to win a true understanding of the atomic world. We were just not worthy.

Neither Planck nor Einstein would ever achieve comfort with quantum mechanics in this form, despite its eventual dominance. Physicist A. Douglas Stone recently relayed Einstein's lifetime toil with quantum with a whiff of Dr. Frankenstein's tragedy, in which Einstein eventually viewed a monster he had helped create with a mix of sympathy, regret, and horror.[24] Hoping to tweak this monster for the better, Einstein jousted with the new quantum mechanics for many years, proposing "gotcha" types of paradoxes to the new quantum apostles. But the paradoxes, after some intense work by Bohr or Born, were always resolved to the vindication of quantum mechanics, and Einstein would slink back to his mental workshop, determined to build a better trap.

In the most fascinating and still relevant example, Einstein and his co-authors proposed a ridiculous outcome from the Copenhagen interpretation. By considering two quantum particles "entangled" at the start of a thought experiment, and letting them fly off in opposite directions, Einstein showed that quantum mechanics would lead to faster-than-light communication. Since this was forbidden by the theory of relativity, something was amiss with either the universe or the new quantum mechanics.

Entangled particles are like two travelers who share and pack a small suitcase, with just one white shirt and one blue shirt. Imagine the travelers become separated but, as quantum particles, they don't really have to determine an outfit until a human spies one of them. When a human observes one particle, the measured particle must *immediately* call its partner and say, "Hey, I'm wearing white today. You've gotta wear blue." Quantum mechanics requires a phone call so immediate that it must travel faster than the speed of light.

Einstein called this conundrum "spooky action at a distance," and the EPR paradox (for authors Einstein, Podolsky, and Rosen), would either refute the Copenhagen interpretation or open an entirely new concept in physics. In time, experiments found that quantum action can indeed get spooky at a distance; entangled particles *do* somehow communicate immediately. Today, some researchers are even exploiting the effect to attempt commercial message encryption. The reality of this particular Einstein relic cracks *another* of our most comfortable bedrock human assumptions, locality. In shorthand: To make a dent in something, you must be standing near it or throw something or beam something at it. Interested readers can pursue the subsequent Bell's theorem of 1964 for further blowing of mind and for further support of quantum mechanics, despite its cruel and steady march away from our intuition.[25] Suffice it to say, Einstein's pithy formulation of the paradox undercuts his own claim that he was becoming more "blockheaded" with age.

"Quantum mechanics is certainly imposing," Einstein once wrote to Max Born, setting up one of his most oft-quoted phrases. "But an inner voice tells me that it is not yet the real thing. The theory says a lot, but it does not really bring us any closer to the secrets of the Old One. I, at any rate, am convinced that He does not play dice."[26]

Planck and Einstein had both come to hold sacred reverence for the concept of causality. In Planck's words, causality is, "the fact that natural phenomena invariably occur according to the rigid sequence of cause and effect. This is an indispensable postulate of all scientific research."[27] But some believed (and believe, to varying degrees) that the probabilistic aspect of quantum mechanics deals a great blow to causality. Planck and Einstein trusted a universe where experiment A provided result B, reliably and definitely. But Heisenberg rejected such outdated notions. "When one wishes to calculate 'the future' from 'the present,'" he said, "one can only get statistical results."[28] Planck saw the younger generation as resigning themselves to ignorance,[29] and he must have heard in their statements echoes from Ernst Mach and the positivists.

Planck was 69 when he first confronted the Copenhagen interpretation. Two generations removed, he didn't debate the young guns like Einstein did. Based on his writings, he came to accept the physics of quantum mechanics, showing one last burst of flexibility. But he fought to maintain his old comforts like causality via philosophical maneuvers. Copenhagen aggravated his old wounds from Mach, and he began a series of philosophical essays, intending to throw a life preserver to his beloved notion of causality.

Planck had what a physicist philosopher today would call a strong or strict version of causality, with little room for nuance: The universe operated via an irreversible march from known cause to certain effect. And at least to my reading, he sometimes conflated causality with the related but even more stringent idea of determinism. In twin essays on "Causation and Free Will" he confronted a very old dilemma: If we support an exact and predictable scientific world view, including the molecules in our brains, then how can we believe that our impulses are anything but pre-determined results? He wrote his essays, "to keep the lines of communication clear between serious science and the seriously thinking public." And he wanted to counteract those who claimed causality was not relevant to quantum physics, an idea that had been "exploited by popularizers."[30]

First, he boldly summarized the "failed" efforts from millennia of philosophers to reconcile a clearly causal universe with the very "dignity of man." Planck wondered, "how far science can help us out of the obscure wood wherein philosophy has lost its way." In the end, he painted a fully causal universe, except for "one single point . . . the individual ego." And science could never move beyond this point, because, "the observing subject would also be the object of research. And that is impossible; for no eye can see itself." In these essays written circa 1930, Planck had his causal cake but enjoyed a cup of free will as well.[31]

Reactions to Planck's earnest offerings were mostly negative, and among philosophers, even dismissive. Heisenberg wrote respectfully of Planck's efforts, although he noted "little practical value." Physicist Wolfgang Pauli, renowned for his blunt ways, labeled Planck's thinking "sloppy."[32]

But Planck kept writing and speaking philosophically. Over the next 10 years, his positions softened somewhat. His essays "The Meaning and Exact Limits of Science" and "The Concept of Causality in Physics," published posthumously, simultaneously move the goalposts of science and paint any refutation of causality as a misunderstanding. The goal of science was, he wrote, "the creation of a world picture, with real elements which no longer require an improvement, and therefore represent the ultimate reality."[33] And yet, "the introduction of the elementary quantum of action destroyed this hope at one blow and for good."[34] Even if Planck was no happier with the new physics than Einstein, he may have been more comfortable shifting his focus from the destination to the journey itself. "To some extent it is unsatisfactory but on the other hand it is proper and gratifying, for we will never come to the end, to finality. Scientific work will never stop, and it would be terrible if it did. ... In science rest is stagnation, rest is death."[35]

Perhaps the best way to summarize Planck's final stance on quantum mechanics would be this: He accepted the theory, equation by equation, but saw an explosion of falsehoods whenever someone applied it carelessly. He deplored a practice that still lingers: the vague application of quantum mechanics to biology and psychology. This seductive mixing began almost immediately after Heisenberg's 1927 paper on uncertainty. Some conspicuously talented physicists like Pascual Jordan and Wolfgang Pauli were notable perpetrators, using quantum theory to support, respectively, the psychologies of Freud and Jung.[36]

Planck especially cringed when others rushed to connect quantum ideas to philosophical conclusions. He bemoaned what he saw as unnecessary philosophical hand wringing over causality. For him, that fundamental philosophical pillar survived the quantum revolution with just a few bumps, scrapes, and clarifications. In this new scientific world, he thought those working at the philosophical extremes were doomed to frustration. Those obsessed with finding "a rule behind every irregularity" and likewise those who threw up their hands and saw "nature ruled exclusively by statistics," were all being too intellectually rigid. "The law of causality is neither true nor false," he wrote late in life. "It is rather ... a signpost—and in my opinion, our most valuable signpost—to help us find our bearings in a bewildering maze of occurrences."[37] In short, what would we do without it? Why would elderly Max put a needle to a phonograph record if he didn't believe that action would cause a soothing Schumann arrangement to fill the room?

In late 1944, Planck's chamber for such thoughts, his meditative library, was in ashes. But the bewildering world brought another random-looking occurrence, a surprising statistical blip. On November 9, the Otto Wolff Company received a letter from the offices of Heinrich Himmler, asking the company to inform Max Planck that his plea had not gone unnoticed and that Erwin's sentence was, for now, suspended. The SS Reichsführer could justify converting the death sentence to life in prison.[38]

Nelly received this incredible news first and caught Max briefly by phone. In a follow-up letter, he wondered if his cries of celebration could be heard all the way from Rogätz to Berlin. He wrote that life itself felt restarted. He immediately thanked Himmler for giving the family a reason to hope.[39] The Plancks received no further news in 1944, and Max's New Year's card to his friend and biographer Hans Hartmann underlines his state of mind and the state of Germany. He thanks the Hartmanns warmly for sending a letter, and he sympathizes with their woes: a younger son at risk in war; an older son facing grave injuries; and a damaged, barely livable home. He thanks them for asking after his granddaughter Grete Marie, saying she had fled Heidelberg, "with her four children helter skelter, overnight," presumably in the midst of the ongoing air raids. Then he turns to Erwin. "The verdict hasn't been carried out yet, and we still have some hope that the pardon will happen, given support from influential places. But I still feel since the verdict was so final that the sword of Damocles hovers over us constantly."[40] He knew Himmler could be overruled, and Max Planck had already measured the Führer's chaotic temperament in close quarters.

14

January 1945

Nazi officials at the Plötzensee Prison mailed invoices to prisoners' families. Dreaded news often arrived in the form of a typewritten bill. A prisoner's family owed 1.50 reichsmarks per day of imprisonment, 12 pfennigs to cover the invoice postage itself, and for a prisoner's execution, 300 reichsmarks.[1]

But by mid-January of 1945, the normal systems were breaking down. Hitler had retreated to his Berlin bunker, and his last gambits were failing in the east and the west. Soviet armies leapt across Poland, and the Allies leveled the German "bulge" near Luxembourg. Himmler ordered the evacuation of the Auschwitz concentration camp in the east, choosing a death march for his prisoners versus impending liberation.

Nelly Planck heard her most dreaded news through unofficial channels and was left to share it with Max Planck. Travel in and out of Berlin was nearly impossible in early 1945, requiring express Reich approval, but Nelly had to tell Erwin's father in person. After failing to get a trip approved, Nelly turned to a car from a neutral state (presumably, for instance, from the Swedish or Swiss embassy), and they conveyed her to Rogätz where she huddled with Max and Marga. On January 23, 1945, the Gestapo had executed Erwin Planck by hanging.[2]

Himmler's earlier gesture to the Plancks had been just a random fluctuation of the Nazi government. "No one understands it," Marga wrote to Max von Laue.[3] The decision was especially shocking following a Reich letter to the Plancks, just days earlier, promising Erwin's imminent pardon. "It all happened suddenly and in secret," Marga wrote to Planck's niece in Göttingen. "The very moment we were certain of his amnesty, the verdict was enforced. It is terrible and we cannot believe it. Himmler is said to be on the Eastern front, but supposedly Hitler was in Berlin when they handed him the lists and he made the orders." In execution, Erwin joined the sons of other Grunewald academic families, like Ernst von Harnack, and Dietrich Bonhöffer. "Erwin was reportedly very composed," Marga concluded in her letter to Göttingen. "But one shouldn't dwell on it. How must he have felt?"[4]

"He was a precious part of my being," Planck wrote to relatives, "he was my sunshine, my pride, my hope."[5] He even shared this pain outside his family. "My grief cannot be expressed in words," he wrote to his colleague Arnold Sommerfeld.[6]

His letters show his characteristic optimism in tatters, but "if there is consolation anywhere, it is in the Eternal," he wrote to one friend. "God protect and strengthen you for everything that still may come before this insanity in which we're forced to live reaches its end."[7]

The last of Marie's blood and the last of Planck's 1905-era home died with Erwin. Contemplating the gaping tragedy of losing a wife and four children is a bit like standing on the rim of Arizona's Grand Canyon; most of us can't hope to really absorb or understand it. But Planck's tally matches those of many Europeans who spanned both wars. If a family had four children born between 1885 and 1895, most were unlikely to ever see 1946. (With life expectancies in the late nineteenth century falling south of 50 years, Max Planck was, speaking in cold statistics, more of an anomaly than any of his children.) Although fate may have been cruel to Planck, it was also emblematic of his time and place—the curve of his life traces that of his homeland.

As Marga's letter suggested, she and Max wondered if Himmler's absence and Hitler's presence in Berlin tipped the final scales against Erwin. Planck's own face off with the Führer must have weighed on him then. Did Hitler recall it with the same clarity that Max always would? Was Erwin's execution the last word of their 1933 confrontation? Shortly after Erwin's death and not long before his own, Planck was asked to recount that meeting and Hitler's eruption. Too feeble to write, Max had Marga transcribe his account for the official record.

The great academic turbulence following Hitler's rise to power framed the meeting. The first and most public fallout involved Einstein. In the spring of 1933, Planck mailed Einstein one last chilly letter, and the Prussian Academy of Sciences voted to expel their most famous member. Einstein mailed one last sad but not accusatory letter to Planck, and in a most prescient line, he described what he saw as a "war of annihilation against my defenseless Jewish brothers."[8]

The day after Einstein wrote that letter, the Nazis unveiled the new Law for the Restoration of Professional Civil Service, requiring civil servants, including all university employees and teachers, to verify their Aryan, non-Jewish ancestry. The first wave of dismissals shed about 15% of all university instructors and professors across Germany. One in ten members of the German Physical Society had to leave their posts or left voluntarily. Many left Germany altogether, with two in three German Jews emigrating by 1938. The government frequently back-filled vacated academic slots with half-competent or totally incompetent party members.[9]

The Nazi's second blow to German physics was their choking the incoming stream of students. Starting in the 1920s, the Nazi party made great inroads with young males, and the majority of college-age boys aligned with the cause as early as 1931. The ongoing propaganda against "Jewish Physics" had pushed students away from physics classes, and in the same month of the new Civil Service law, the Nazis also unveiled the Law Against Overcrowding in Universities, greatly limiting the number of Jewish students. During the first seven years of Hitler's rule, physics student numbers dropped 75%, and mathematics, the closest ally of physics, lost 95% of its students.[10] Planck and his colleagues watched German physics step off a cliff, although its worldwide reputation would linger for some time.

Some physicists left voluntarily. The Jewish physicist James Franck qualified for the veterans' loophole in the Civil Service law, but he made one of the more brave and public gestures. He not only resigned his post, but also sent an excerpt of his resignation letter to the press, decrying that German Jews were now treated as enemies, that they were to raise their children as inferiors. A newspaper column supported Franck and claimed "the loss to science is beyond estimation," but the column's author remained anonymous, marking the rising tide of fear.[11] Franck's gesture earned private support from friends like Edith Hahn, wife of Otto. "You both did something," she said, writing to James and his wife Ingrid. "This is what is so marvelous. . . . I hope the whole world will react to it."[12]

Otto Hahn himself could count on his ongoing work with Lise Meitner, who opted to stay. As an Austrian citizen, Lise was exempt from the laws, and perhaps as a woman bucking the odds in science, she saw the new government as just another nuisance.[13] Nevertheless, the University of Berlin canceled her spring 1933 course, team-taught with the Jewish physicist Leo Szilard. Szilard fled to London and eventually became a prime mover in America's Manhattan Project. Left behind, Lise hoped to teach in the

Figure 14.1. President von Hindenburg and Chancellor Hitler on May 1, 1933, about two weeks before Hitler's meeting with Max Planck.
Courtesy the Bundesarchiv, Photograph 102-14569 by Georg Pahl.

following semester, but by then the Nazis removed her professor's title and stripped her of all teaching rights. Hahn and Planck's strong letters of protest had no effect. Yet they assured Meitner that they would be able to protect her from any further Nazi outrage.

One case ruffled Planck enough to bring it directly to Hitler's attention. The Jewish chemist Fritz Haber directed the Kaiser Wilhelm Institute of Physical Chemistry. He was also a decorated scientist who had toiled for the Fatherland in World War I. In resigning his post and waiving his veteran's exemption from the Civil Service laws, he said he refused to consider racial factors in staffing his institute. If his Jewish colleagues had to leave, then he would leave with them.[14] Planck and Haber had traveled to receive their Nobel Prizes together, and they had brainstormed the *Notgemeinschaft* to keep German science afloat during the great inflation. Watching the new government chase away someone of Haber's caliber and service was just too wrongheaded to Planck.

So when he received a birthday note from the new Reich's Chancellor on or around April 23, 1933, he sent a friendly reply, requesting a meeting. He wrote that he wanted to make sure he aligned science to promote the greatest good of the Fatherland. Hitler accepted the meeting, set for 11:00 a.m. on the morning of Tuesday, May 16 (Figure 14.1). As Planck awaited the audience and considered his strategy, he would have heard about the enormous bonfire adjacent to his long-time university office. On May 10, the SS and Josef Goebbels organized a book burning at the University of Berlin. Nazi-sympathetic students carried 20,000 books from the shelves of the university library, piled the works of Thomas Mann, Karl Marx, and hundreds of others in nearby Opernplatz, and set them alight. To employ Planck's style of Prussian understatement: This was not an encouraging sign of academic support.

Accounts of the Planck–Hitler meeting all trace to Planck himself: either via those who spoke with him afterward or from the account typed by Marga in 1947. Here, we follow the reconstruction offered by science historian Kristie Macrakis in *Surviving the Swastika*.[15]

After exchanging pleasantries, Planck began by assuring Hitler that the new Reich could count on the Kaiser Wilhelm Society's (KWG) unwavering support in rebuilding the Fatherland. He requested ongoing monetary support, and Hitler told him not to worry. Planck then brought up Fritz Haber as an example of a Jewish colleague who had always focused on the greater good and glory of Germany. The new Civil Service laws could hurt the Fatherland in special cases like Haber's.

Hitler reacted strongly, claiming he had "nothing against the Jews," being focused instead on the real enemy, communists. The problem was, he said, that all the Jews happened to also be communists.

Planck tried to delicately redirect, saying there were different kinds of Jews. He suggested distinguishing western Jews from eastern Jews, where the former made great efforts to assimilate and contribute to Germany. The eastern/western rhetoric typically implied separating Jews by their perceived utility.[16] (We can't know the extent to which Planck actually believed this. He certainly used such language to win sympathy for his mission, but the underlying sentiment was probably not repulsive to him either. As he had written to Einstein just two months earlier, he saw that Judaism and Nazism couldn't co-exist, and he deplored them both anyway.)[17] Planck told Hitler it was self-destructive to chase the best of Western

Jews from science, not only as a loss to the Fatherland, but also as a benefit to other countries and potential enemies.

Whatever Planck's strategy, Hitler was having none of it, and he grew angry. The problem with Jews, he said, was that where there was one, you would soon find more and more. "They clump together like burrs," he said, no doubt including the many Jews employed by Haber.

As Hitler raised his voice, Planck watched his most reliable instrument, reasonable discourse, being unplugged and tossed aside. Maybe he even apologized for upsetting his host.

"People say I suffer occasionally from nervous debility," Hitler said then. "This is slander. I have nerves of steel." According to Planck's account, Hitler then "hit himself hard on the knee, spoke faster and faster, and flew into such a rage that I could only remain silent and withdraw."[18]

Hitler's nervous condition is well documented, and many link it to his experiences in World War I. Like countless other soldiers, including Planck's sons, Hitler endured merciless shelling and horrific sights of carnage. In a trench letter of 1918, he asked his correspondent to, "forgive my poor hand. I am very nervous right now. Day after day we are under heavy artillery fire . . . and that is bound to ruin even the strongest nerves."[19] He also described himself as having survived a mustard gas attack in October of 1918. "As early as midnight, a number of us passed out, a few of our comrades forever. Toward morning I, too, was seized with pain which grew worse. . . . A few hours later, my eyes had turned into glowing coals; it had grown dark around me."[20] The attack landed him a long hospital recovery, where he experienced news of Germany's surrender as the worst wound yet. A recent retrospective psychiatric evaluation showed that Hitler's overall symptoms fit the signs of post-traumatic stress disorder.[21]

The author Thomas Mann, shortly after learning of his torched library volumes, heard of Planck's meeting through a mutual friend. His third-hand account relays Planck's reaction to Hitler's outburst through Mann's own Nazi-loathing filter. Planck listening to the Chancellor was, "like listening to an old peasant woman gabbling on about mathematics, the man's low-level, ill-educated reliance on obsessive ideas; more hopeless than anything the illustrious scientist and thinker had ever heard in his entire life."[22]

Just one week after meeting Hitler, Planck gave his infamous address to the KWG saying that no man would be "allowed to stand, rifle at rest." He read a message on behalf of the KWG to Hitler, saying that German science was "ready to cooperate joyously in the reconstruction of the new national state." At the same meeting, however, Planck stated for KWG records that he had been on vacation when they drafted the strident condemnation of Einstein.[23] This tiny nod to friendship complete, he pivoted and removed 12 non-Aryans from the society's senate.[24]

The next day, Planck penned a note to his former student Gabriele Rabel. "In these times," he told her, "you count yourself lucky to retain any optimism or lightness."[25]

And whither Fritz Haber? When the Reich failed to even blink at his resignation, Haber knew he had to leave the country. In late July, he visited the American ambassador to Germany, William Dodd. Distraught, in poor health, and trembling in fear at his predicament, Haber asked Dodd if America might welcome an accomplished scientist, but the bumbling Dodd mistakenly told Haber it was impossible, because the German emigration quota was full. Haber thanked him and asked that he not mention their meeting to anyone.[26] He later made his way to England, and Einstein was particularly cruel to the assimilation-obsessed Haber at this point. "I am delighted to learn that your love for the blond beast has somewhat cooled," he wrote.[27] But Haber wouldn't serve as Einstein's moral punching bag for long—he was dead of a heart ailment within months. His obvious love for his country lives on poignantly. As he said a last goodbye to his homeland, he donated his original ammonia synthesis apparatus—that which gave humanity widely available fertilizer—to Munich's Deutsches Museum.[28]

Meanwhile, as Planck filed KWG reports and requested exceptions to Nazi policies, he watched continuing losses for German science.[29] Max Born, another Jewish physicist who could claim the veteran exemption, seemed poised to leave. Unknown to Planck, Born had maintained correspondence with Einstein, who warned Born in March that the Nazis were, "moving in a direction that will become increasingly destructive."[30] Heisenberg made appeals for Born to stay, strangely citing Planck's meeting with Hitler and claiming that Hitler promised no further Nazi meddling in science. Born fortunately rolled his eyes at this, forwarded the ridiculous claim to a friend, and packed his bags for Cambridge.[31]

As noted previously, some saw Planck's acquiescence as his best available route to protect people and change the situation. But, as the science historian Michael Eckert points out, Planck was also probably reluctant to step away from his accustomed access to power.[32] Whatever the mix of motivations, it wouldn't take long for him to see what little influence or power he held with the new regime, and the records of the KWG show only meager rewards for his accommodation.[33] In October, Lotte Warburg recorded an unflattering vision in her diary. "Planck has become a stone-aged man, stooped, pitiful. I saw him shuffling through the park, untidy and unkempt. I gave him all the news about [firings at] the universities. He knew nothing about it. He said that no one asked him about things anymore, that science was no longer worth anything."[34]

Planck's interaction with and reaction to the Führer's rage can be understood as more than a pragmatic attempt to save science. He had relied on being well liked starting from his earliest conscious moments in the family home and throughout his schooling. He continued as a young man to seek the good graces of his superiors in physics, showing up on Professor Clausius's doorstep, yearning for approval. Though many academics dislike authority, it pulled on Planck like a magnet. And while he could be direct and stand his ground, especially scientifically, Planck also could suffer deep wounds from the opinions of reputable colleagues.

His scientific autobiography reveals a great deal about Planck the human being. The document exhibits his normal clarity and humility, but it spans only 38 pages, leaving out countless conversations, side projects, and technical wrangles. Although he sticks to purely scientific exploits, with no mention of life beyond work, he tells the stories with a disarmingly personal tone. Hear him speak to his first positive and meaningful scientific feedback, arising from none other than Hermann von Helmholtz, the leading European scientist during Planck's early career—an ultimate authority figure.

> When during a conversation he would look at me with those calm, searching, penetrating yet so benign eyes, I would be overwhelmed by a feeling of filial trust and devotion, and I would feel that I could confide in him, without reservation, everything that I had on my mind, knowing that I would find him a fair and tolerant judge; and a single word of approval, let alone praise, from his lips would make me as happy as any worldly triumph.[35]

Such gushing was not characteristic of Planck's writing.

Meanwhile, despite the pared length of this scientific story, he made room for former slights that he felt decades later. Shortly before starting his post in Berlin in 1889, he had published some interesting results in electrochemistry, but that work "unfortunately got me into an unpleasant conflict." The antagonist was Svante Arrhenius, a Swedish scientist of Planck's generation who had recently introduced the idea of positive and negative ions to explain electricity in fluids. "Arrhenius challenged, in a rather unfriendly manner, the admissibility of my arguments."[36] Arrhenius later became a true Planck champion, publicly declaring Planck's genius and lobbying for his recognition.[37] Yet we only see the "unfriendly" Arrhenius mentioned in the *Scientific Autobiography*, penned more than 50 years after their minor dustup. Planck was even more wounded by the mercurial Ludwig Boltzmann. After Planck's star pupil, Ernst Zermelo, criticized a point of Boltzmann's, the Austrian replied "in a tone of biting sarcasm, which was meant partly for me, too.... This was how Boltzmann assumed that ill-tempered tone which he continued to exhibit toward me, on later occasions as well, both in his publications and in our personal correspondence."[38] In Planck's account, there's nearly as much of Boltzmann's hurtful manner as there is of his science.

In summary, Planck exhibited more of a "Can't we all get along?" persona than someone who could sit comfortably with high-level confrontation. If so, the effect of Hitler's outburst, which would have disturbed anyone, could easily have led to the image that Lotte Warburg painted of Planck months later, "shuffling ... untidy and unkempt," if she is to be believed.[39] A 75-year-old man with Planck's personal settings, his lifelong allegiance to the state, and his pragmatic survey of what a physicist calls "boundary conditions" would leave one conceivable path: working within the new system, however deranged, and trying to make a positive difference, however small.

At the end of 12 years on that path, he found few if any Pyrrhic victories. Instead, he endured the steady loss of friends and colleagues. And finally he faced the news of Erwin's death.

The stories that now trickled to Nelly, Marga, and Max painted Erwin as brave by comparison. Fellow prisoner Otto Gessler was released shortly after Erwin's execution, and he wrote to Nelly. "The memory of your husband is held in the highest esteem," by fellow prisoners. "He was a man of honor as few were."[40] Erwin's friend Georg Thomas had occupied a cell next to him in late 1944. Thomas reported that the Gestapo interrogated and tortured Erwin, demanding a statement against Thomas, but Erwin wouldn't yield, declaring to the last that his friend had nothing against Hitler. Meanwhile, Erwin comforted Thomas in the midst of his friend's ongoing heart ailments.[41] Erwin's last weeks can also be glimpsed through fellow resistance member Fabian von Schlabrendorff, who miraculously survived the Nazi era and wrote of his experiences. After describing the four-stage Gestapo torture techniques, von Schlabrendorff writes that he, Erwin, and others, "could endure far more than we ever believed possible," calling upon, "the great polar forces": love of their cause and hate of the regime.[42] And one of Erwin's closest childhood friends Helmuth Rhenius was interrogated with Erwin, as the Gestapo confronted them with conflicting testimony. He relayed a noble portrait of Erwin's last days.

> The leg irons only allowed him to walk slowly. He went to the middle of the room where a bright lamp, dangling right over him, drew sharp shadows on his haggard face. He greeted the interviewing Gestapo agents with a short nod.... He was standing ... cuffed at hands and feet, but straight and with his head high, with an expression and posture of a man who had overcome all the agony.[43]

15

April 1945

Shortly after Erwin's execution, Allied bombing put a permanent end to the People's Court. Hitting it mid-session, the airstrike destroyed the courthouse, and an enormous beam cascaded through the shelter beneath, killing Judge Roland Freisler. He died with prisoner files still in his grip.[1] The Plancks received little consolation from this news. Nor were they cheered by the waves of written condolences, sharing their own fresh tragedies: destroyed apartments, missing relatives, dead soldiers, and the deprivations of a nation in collapse.

Planck spent time at the guesthouse piano playing Erwin's favorite songs. In writing to his nephew, he mistakenly wrote Erwin as his signature. "I fight every day anew to gain the strength to accept this twist of fate," he wrote. "Because with every new morning, it overcomes me like a new blow, paralyzing me and clouding my clear mind, and it will take a long time until I'll be back in mental balance."[2]

In early April, American troops swept eastward toward the Elbe River, while Soviet armies took Vienna and moved rapidly on Berlin. Marga was terrified at the increasing military presence around them. An officer told the Plancks that the area was ideal for an enemy's aerial troop landing. And the medieval tower adjacent to the guesthouse would make a good perch for German artillery, anchoring their defenses. As German soldiers swarmed into Rogätz, Marga wrote a hurried letter to a relative in Göttingen, asking if she and Max could stay there in an emergency.[3]

As the Americans approached, General Dwight D. Eisenhower called for German surrender, but Hitler ignored it. "Never will we repeat the mistake of 1918," he had declared in 1943, "laying down our arms at a quarter to midnight."[4]

Meanwhile, the world began to witness the scope of Nazi atrocities as the British liberated the Bergen-Belsen concentration camp, piled with unburied bodies, and the Americans found Buchenwald. On April 12 (the day American president Franklin Roosevelt died), journalist Edward R. Murrow entered Buchenwald, and his CBS broadcast became a particular touchstone of the Holocaust.[5] He began by warning listeners that his report "would not be pleasant listening." After talking with several prisoners at the gate, he approached one of barracks. "When I entered, men crowded around, tried to lift me to their shoulders. They were too weak. Many of them could not get out of bed. I was told that this building had once stabled 80 horses. There were 1,200 men in it, five to a bunk." Driven outside by a stench "beyond all description," he then watched men in their final moments, slumping and expiring before him as others crawled toward the latrines. General Eisenhower began shuttling large tours of German citizens through the camps, where many broke down in tears or fainted.[6]

Max and Marga wouldn't have heard the news. Rogätz became a war zone by April 9, and the Plancks fled their once quiet refuge amid approaching gunfire and shelling. They crossed the Elbe River and headed east. On April 13, American troops claimed Rogätz, but intense fighting continued in the area. Now the Plancks tried to locate primitive shelter, barns and sheds, while navigating what Marga later described to Lise Meitner as a "combat zone" filled with the "last but not the least horrors of the war."[7] Only mortal danger could compel them to move, as Planck's curled and rigid back rendered him nearly immobile. "The very worst was the frightful suffering that Uncle Max had to bear," Marga later wrote. "He often screamed from the pain."[8]

There was no returning to Rogätz, even if they'd wanted to surrender to the Americans. German troops continued to launch counter attacks, seeking to retake the town.[9]

As Planck fled the battle, did he glimpse an American face bobbing beneath a green helmet? Would he recall his former encounters and fond impressions? The happiest time of his life's long arc included his only trip to America. In early 1909, all the dear women in his life were still with him: his mother, his first wife Marie, his friend and colleague Lise Meitner, and his musical twins Grete and Emma.

His lectures now attracted 10 times the students that his first Berlin lectures had 20 years earlier. He was only a few months from finally meeting his pen pal, Albert Einstein, at a Salzburg conference. In 1909, Planck foresaw the two of them leading the way in building a single theoretical structure to unite all of physics.

Columbia University invited Planck to lecture on the state of theoretical physics. He looked forward to taking such a trip with Marie, but her health steadily worsened as the trip approached. She started with a cough that winter, and "gradually a fever came to her," he wrote in the letter diary, and "she is confined to bed, with a left-side pneumonia." He took her to a sanatorium in Ebersteinburg, "where she now hopefully finally goes to find healing with careful and expert care. The major regret is that she cannot accompany me on a journey . . . to New York." Instead of traveling alone, he elected to take one of his daughters in Marie's place, choosing the "older" Emma, and presenting Grete with a harmonium as a consolation prize.[10]

As far as we know, this marked Planck's only overseas trip, and if he recalled the stories from his mentor Hermann von Helmholtz, he followed those footsteps with a mix of excitement and trepidation. Helmholtz and his wife had traveled to see the Chicago World's Fair of 1893 and, in particular, to attend the International Electrical Congress (along with Thomas Edison and others). At the time, Helmholtz was nearly without peer as an international scientific figure, and America greeted him with much fanfare.[11] He met with President Grover Cleveland, and American elites queued for a handshake as the famous scientist toured major cities. But on the return ocean voyage, he'd tumbled down a flight of stairs, striking his head. Everything healed externally, but the 71-year-old never regained his former energy and concentration. One year later, back in Berlin with Planck, a cerebral hemorrhage sent him to his deathbed. So while Planck may have associated the trip to America with the great Helmholtz's demise, he must also have recalled the man's enthusiasm. "I am convinced that America represents the future of civilized humanity," Helmholtz wrote to a friend in 1893, "while in Europe we have only chaos or the supremacy of Russia to look forward to."[12]

Max and Emma first stopped in Washington, DC, and he wrote Max von Laue a postcard featuring the White House (bereft of trees save a few scrawny palms), mentioning that New York was next.[13] Planck relayed to his friend Willie Wien that he'd found "vigor and hospitality" in America and that he cherished Emma's company, as they shared impressions and navigated busy itineraries.[14]

The New York of 1909 boasted twice Berlin's population and would have surprised any German tourist with its young skyscrapers. Several buildings already rose more than 20 stories and the new Singer Tower was the world's tallest, rising over 600 feet, with 47 floors. One assumes Max and Emma would make the short walk from Columbia to the city's Central Park, but they might have found less relief there than in their native Grunewald paths. Automobiles were now allowed to jostle with horse-drawn and goat-drawn carriages throughout the park.[15]

When Max began his eight Columbia lectures, he did so as the most renowned and respected theoretical physicist in the world. As one generation faded, he transitioned to the old guard. His mentor Helmholtz had passed 15 years earlier, Boltzmann had recently taken his own life, and Einstein, despite his miracle year of 1905, didn't yet enjoy Planck's international reputation. In 1909, relativity still seemed radical, and very few took Einstein's "light quanta" idea seriously. To an American audience, Planck was the most brilliant physicist from the most prestigious scientific nation, and they eagerly awaited his lectures, which he naturally delivered in German, the language of physics.[16]

He thanked the president of Columbia and began with a humble caveat, saying he couldn't possibly discuss all of theoretical physics. Instead, he would hit the primary topics of his greatest interest and involvement. From the outset, he warned against either underestimating the success of the topic, pointing to recent technologies like "wireless telegraphy" and "aerial navigation," or overestimating its prospects. For instance, he dismissed recent claims that physics would soon describe both the inner workings of atoms and also "the laws of mental life."[17]

He built quickly to one of the best encapsulations of his ultimate vision.

> In short, we may say that the characteristic feature of the entire previous development of theoretical physics is a definite elimination from all physical ideas of the anthropomorphic elements. . . . Now, what are the great advantages to be gained through such a real obliteration of personality? . . . The result is nothing more than the attainment of unity and compactness in our system of theoretical physics, and, in fact, the unity of the system, not only in relation to all of its details, but also in relation to physicists of all places, all times, all peoples, and all cultures.

This is the early talk of the "unified field theory" that later haunted Einstein's work and compels theorists to this day. But Planck wasn't done with the topic of transcendence.

> Certainly the system of theoretical physics should be adequate, not only for the inhabitants of this earth, but also for the inhabitants of other heavenly

> bodies. Whether the inhabitants of Mars, in case such actually exist, have eyes and ears like our own, we do not know—it is quite improbable, but that they, in so far as they possess the necessary intelligence, recognize the law of gravitation and the principle of energy, most physicists would hold as self evident.[18]

Our red neighbor was a hot topic in 1909, especially in the United States. The polymath Percival Lowell had turned his thick bankroll to studying astronomy at about the same time Planck turned to black-body radiation. Lowell built his observatory just outside Flagstaff, Arizona. (The installation functions to this day and offers a fascinating tour.) After many years of work, he published *Mars and Its Canals* in 1906, followed by *Mars as the Abode of Life* in 1908. Lowell reheated and popularized the notion that the visible striations on Mars were actually the canals of an intelligent civilization. But the public welcomed these books much more than did the scientific community. Soon enough, other more reputable astronomers, using a more powerful telescope, said the "canals" were probably signs of natural erosion on the barren Martian surface. So although Planck played to his audience a bit with the Mars reference, he probably gave his fellow physicists a knowing wink as well. Not catching the wink was his philosophical nemesis Ernst Mach, who ridiculed Planck that year for referencing the science of aliens.[19]

In the meat of his first Columbia lecture, Planck staked his central pillar of physical thought: entropy. By differentiating the majority of all physical interactions, in which entropy would increase, from the very rare cases, in which entropy might remain constant, he pointed to the absolute but still confounding arrow of time. "I express it as my opinion that in the theoretical physics of the future the first and most important differentiations of all physical processes will be into reversible and irreversible processes."[20]

He delivered one lecture per day, and after the first week, Planck had come to his own dear baby: the theory of "heat radiation," explaining the black-body experiments. The thoughtful progression of Planck's lectures showed a master architect's eye at its best. Moving from the existence of atoms to Boltzmann's statistical treatment of atoms and the great conceptual power of entropy, he set up the black-body problem as a natural or even necessary destination.[21] Using some of Boltzmann's tools, as in his 1900 work, Planck derived his famous formula, but even in 1909, he concluded by shaking his head. He confessed that, "the physical meaning of the universal constant h remains quite unexplained." And although he admitted that all existing classical theories "suffer from an incompleteness which demands a modification," he was not yet ready to embrace the "radical" ideas of Einstein and others, in which light itself flew about in particle-like quanta. He modestly suggested, "It is not necessary to proceed in so revolutionary a manner" and proposed that the quantum of energy existed only between nearby components of the black body, with no direct relevance to light itself.[22] So nine years after his discovery, the public Planck still sought to chain it to the fence separating matter from light, unwilling to let the quantum roam off leash in either region.

In his eighth and final Columbia lecture, Planck turned to the theory of relativity as the cutting edge of physics. Instead of diving right into Einstein's principles, he set out the historical narrative moving through James Maxwell, Heinrich Hertz, and Hendrik Lorentz before placing Albert Einstein as the next logical way-shower, followed closely by Einstein's former mathematics professor Minkowski. Halfway through this lecture, he made a crucial pivot that presages the political dawn of "German" versus "Jewish" physics in the devastating decades to come. Einstein's ideas abolished the idea of absolutely simultaneous events or a universal time for all observers, and Planck admitted this would spawn headaches. Since the human mind was poisoned with preconceptions of time and therefore prone to, "logical mistakes . . . we shall adopt the mathematical method of treatment."[23] From there, his lecture was heavily and beautifully mathematical. We can see where the mathematical bus of Planck and Einstein was leaving the Johannes Starks of the physics world coughing in exhaust fumes.

He concluded this final lecture with an example that could strike one now as esoteric. He considered a black body moving at very high speed through a laboratory. By combining his own work with Einstein's special relativity, he computed the speeding body's thermal radiation spectrum. Especially in 1909, this was not a realistic or useful case, so what was he doing? By intentionally entwining their two greatest creations, he wanted to demonstrate that physics was moving toward a unified picture. But only a few months later, at a Salzburg conference, Planck heard Einstein bluntly declare that Planck's own work meant there was no turning back—classical ideas of light were as good as dead. Though they would come together in Berlin as friends and music partners, their two signature theories eventually went the way of their personal dialogue, separate and irreconcilable.

"Colleagues, ladies and gentlemen, I have arrived at the conclusion of my lectures. I have endeavored to bring before you in bold outline those characteristic advances in the present system of physics which in my opinion are the most important. Another in my place would perhaps have made another and better choice." He promised ongoing work, to leave "lasting results" for the next generations of physicists. "For if, while engaged in body and mind in patient and often modest individual endeavor, one thought strengthens and supports us it is this—that we in physics work not for the day only and for immediate results, but, so to speak, for eternity." He then thanked them for their patience.[24]

Steaming eastward across the Atlantic, the professor and his daughter must have traded impressions of America. In addition to New York and Washington, they had visited "Baltimore, Boston-Cambridge, Ithaca, Niagara Falls. . . . Everywhere the people were lovely and amiable."[25] Max Planck's recollections were indelibly upbeat, even reverential. A few years later, in his first address as rector of the University of Berlin, he cited America's energy and optimism as worthy models for Germany.[26] But when Max and Emma returned to the Planck villa in Grunewald, they found Marie worse than ever, with a fever that never abated. She died of her lung ailments within months, and Max buried her next to her parents in the mountains near Tegernsee. As more Planck funerals quietly queued for the coming decade, and as younger physicists prepared to close the chapter of his intellectual leadership, he left the peak of his long and gracious life in the wake of the passenger liner returning him to Germany.

In 1945, the Americans had come to his Fatherland. They battled retreating Germans through the streets of Rogätz, destroying the town's normal tranquility. They advanced to the western edge of Elbe River, while the German forces regrouped on the opposite bank. The Americans could have crossed the Elbe and moved on Berlin, but General Eisenhower calculated military versus political aims. The Allies had already agreed to a framework partitioning in which the Soviet Union would take the German regions surrounding Berlin. The General saw no use in sacrificing further American lives for the political feat of being first to Berlin, if they then must return to the Elbe anyway. Furthermore, his orders had been to dismantle the German military and then divert resources to the Pacific theater. British Prime Minister Winston Churchill, sensing the iron curtain to come, questioned the strategy of surrendering key cities like Berlin, Prague, and Vienna to the Soviets, but the Americans simply wanted a speedy end to the war's European chapter.[27] Churchill's arguments lost the day, and Eisenhower held fast at the Elbe.

Max and Marga would not have known of Eisenhower's decision. Fighting just to move, they tried to stay safe. What would the Americans do with captured Germans? How long could Max Planck live without shelter in his condition?

Unbeknownst to the Plancks, a new menace grumbled forward from the other direction. The Red Army had overtaken Berlin and moved forward now to meet their American partners at the Elbe, unleashing violence along the way. The Soviet troops looked like the cast of a post-apocalyptic nightmare, as described by author Antony Beevor, "an extraordinary mixture of modern and medieval: tank troops in padded black helmets, Cossack cavalrymen on shaggy mounts with loot strapped to the saddle, lend-lease Studebakers and Dodges towing light field guns, and then a second echelon in horse-drawn carts." To revenge themselves for the brutal German invasion of the Motherland, these troops took to raping women and girls, and shooting any male who dared to interfere. The occupying Red Army raped an estimated two million German women in 1945, regardless of age, pregnancy, or religious garb.[28] As this force moved toward the Elbe, Max passed his eighty-seventh birthday in agony, with Marga comforting him as best she could.

16

May 1945

The journalist Alan Moorehead described Germany near the war's end.

> All around us are things too monstrous to grasp. . . . Ten million people roaming helplessly through the countryside without homes, their relatives lost, and all normal hope gone out of their lives . . . the debris of three-year-old bombings has long since returned to its original dust: locomotives and churches and city halls lie tossed aside in the streets.[1]

Among the homeless citizens were hundreds of thousands of war orphans. And by May, eight million enslaved foreign workers, mainly women and children, started navigating their daunting paths home.[2]

The astronomer turned soldier Gerard Kuiper stood in this aftermath, hardly recognizing the landscapes he'd known in his youth. The 40 year old was a natural for America's top-secret Alsos Mission. Born and raised in the Netherlands, he knew parts of Germany well and spoke the language fluently. So just six years after gaining American citizenship, he had joined the U.S. Army and returned to northern Europe, not that far from his homeland. Now, with the war devolving to a new sort of multinational chaos, Kuiper had a rumor in hand: The scientific treasure known as Max Planck was supposedly alive but on the run near Rogätz, in the Soviet Red Army's grim path.[3]

The Alsos Mission sprouted from the Manhattan Project, and like the secret effort to build an atomic weapon, Alsos combined military leadership with scientific workers. The mission's singular goal was to determine the level of progress and all technical details of Germany's wartime atomic research. The American physicist Samuel Goudsmit served as the scientific lead, but Alsos ultimately answered to General Leslie Groves, who guided the Manhattan Project. Goudsmit relayed the origin of their moniker in his incredibly open recount of the mission. "Secrecy, it was impressed upon us at the outset, was imperative, despite the fact that our code name, Alsos, seemed a giveaway, being the Greek translation for Groves."[4]

In 1943, the small Alsos team of 20 military and scientific people began their work and quickly became a clan. As the months went by, they were given increasingly greater independence to make decisions and set priorities in their chaotic field locations. The aim of Goudsmit, Kuiper, and their Alsos colleagues was to stay just behind the advancing Allied front, ferret out the key German scientists working on atomic weapons, and recover their experiments. Goudsmit, also born in the Netherlands, made an interesting and poignant choice for lead scientist, as he had just lost his parents to the Nazis. "In March, 1943, I had received a farewell letter from my mother and father bearing the address of a Nazi concentration camp. . . . It was the last letter I had ever received from them or ever would."[5] Of all the Alsos scientists, only Goudsmit had been briefed on American's own bomb project.

Most physicists working in America assumed that the Germans had every advantage on the road to an atomic weapon. The once well-earned reputation of German science remained in the minds of most non-German physicists, despite its relative decline since World War I. Moreover, the Germans had pioneered nuclear fission, starting with the Berlin chemist Otto Hahn (and Lise Meitner's critical contributions). Given their long superiority and such a head start, how could Hitler's nation not be ahead of the Allies?

By late 1942, some American physicists battled fears so intense as to approach paranoia.[6] At the time, Axis powers controlled most of Europe, a huge chunk of northern Africa and large parts of southeast Asia. Meanwhile, Allied intelligence revealed the Germans pursuing a well-advanced rocketry program—surely these efforts were meant to deliver some sort of deadly atomic bomb anywhere in the world.

But by late 1944, Goudsmit had found what he most wanted: a non-smoking gun, or in this case a non-glowing pile. In Strasbourg, his team uncovered some innocuous-looking files. To Goudsmit's happy shock, the documents showed that the Germans, "were about as far [along] as we were in 1940, before we had begun any large-scale efforts on the atom bomb at all."[7] Despite his relief, the mission continued (Figure 16.1). From here, Alsos would work on forensic clean up instead of an urgent rush to thwart a German bomb. The Americans prioritized finding the likes of Max von Laue, Otto Hahn, and Werner Heisenberg before the Russians did. Once Alsos had captured them, Goudsmit had to recommend whether

Figure 16.1. Samuel Goudsmit (*left*) and other Alsos members study documents in Stadtilm, Germany, roughly 240 km south of Rogätz, in April 1945.
Courtesy Brookhaven National Laboratory.

each German physicist should be interned or simply let loose, and he said that Otto Hahn and von Laue made for two of his toughest decisions. He understood that von Laue, for one, had never dipped his fingers into the uranium project and had even spoken his mind against the Nazis. But in the final weeks of the war, Goudsmit thought internment was the best option for both of them, if only to make sure they lived to help Germany back to its feet.[8]

When Alsos captured Philipp Lenard, he labeled himself the "greatest scientist" in all of Germany. "Ignore him!" Goudsmit ordered. As he relayed in his memoir, "This, for a Nazi, was a greater punishment than being tried in Nuremberg."[9]

Compared with some of the more straight-laced members of Alsos, Kuiper fancied himself a daring fellow, whether setting his eye to a telescope or a rifle scope. The Alsos plan did *not* include zooming past the front lines and risking interactions with desperate German forces or the Soviet juggernaut. Nor should someone like Kuiper go after a retired and geriatric scientist who had nothing to do with a German bomb program. But he requested permission to go after the legendary Max Planck, and Goudsmit consented. So on May 16, Kuiper grabbed two soldiers, leapt in a jeep, and bounded across the post-apocalyptic landscape toward Rogätz. They made their way past the Elbe River and left the protection of the American forces for a fearful terrain of hiding refugees, stray German army units, and the menacing approach of the Red Army. The latest reports meant time was short. The Russians had already reached the Elbe near Torgau, just some 80 miles upriver from Rogätz.

The rumored information on the Plancks' location proved accurate, and Kuiper found them squeezed into a milk farmer's one-room hut, huddled with the farmer's family over a meager lunch.[10] Kuiper introduced himself and said he would take the Plancks to the American zone. Marga later told von Laue that she could cope with all the many difficult memories but two: the moment they learned of Erwin's death and the anguished cries of her husband as the Americans lifted him into the jeep.[11] And so, just hours ahead of the Red Army, the lone jeep carried Max and Marga to safety, with each bumpy jolt hurting Planck's back anew.

Crossing the Elbe and joining the Americans, the Plancks possessed nothing aside from their lives and their memories. For Planck, this marked an endpoint. Germany had crumbled again into divided segments for the first time since his twelfth birthday. The Empire's 40-year rise and 30-year fall were complete—just like the parallel paths of German science and Planck's own family. The comparison was sadly not lost on him. After his last trip to Berlin, he'd written to von Laue comparing himself physically and intellectually to the broken rubble of their once exhilarating capital.[12] When they arrived in Göttingen, Marga checked Max into a crowded hospital, where he recovered some of his energy over the next several months.

While Alsos left Planck to recoup, they now held a set of German physicists suspected of atomic weapons research. American forces had finally captured Heisenberg after the fall of Munich, and Goudsmit suffered an awkward meeting then. He'd known Heisenberg well before the war, but now he had to let his former conference pal believe that, indeed, Germany was still way ahead of America when it came to nuclear weapons. But after Alsos interviewed the captive German physicists, they became a purely military problem. In the debate concerning what to do with these brainy prisoners, one American general suggested that they should simply execute the physicists while they had the chance, to make sure Germany would never have an atomic bomb. British intelligence officials reacted strongly and suggested an alternate plan: sequester and monitor.[13] Ten key Germany nuclear scientists were then secreted to a comfortable but remote location in rural England, where their conversations were secretly bugged. A team of bilingual soldiers transcribed the conversations in real time, and the top secret files were only made available in 1992. One of the first things the transcribing team caught was Heisenberg scoffing at a younger colleague who wondered if their new residence was bugged. Heisenberg laughed the notion off as impossible, saying their captors were not as sophisticated as the Gestapo.[14]

The physicists were kept at Farm Hall for a total of six long months, beyond the end of the war. Tensions flared from time to time, as the prisoners must have felt like the characters in Jean-Paul Sartre's 1944 play *No Exit*. For months, they were allowed no communication, even with their families. Some of the men pleaded with their captors to allow letters to wives and children or to make sure their families were in the safer American or British zones.

The transcribed conversations provide a unique look at the 10 expressing themselves without censoring their thoughts. The talk was sometimes technical, sometimes personal, and sometimes political. They discussed with great accuracy why the Allies needed to keep them a secret and to keep them away from the Soviets. They mused about their slim options for making a living in postwar Germany, wondering if they could sell their expertise. They foresaw coming tensions, wondering if the next war would be fought between Russia and the United States on German soil.

The atomic blast at Hiroshima came to the attention of Farm Hall at dinner on August 6, 1945. "The impact on the ten scientists was shattering," Goudsmit recalled. "Their whole world collapsed. At one stroke, all their self-confidence was gone and the belief in their own scientific superiority gave way to an intense feeling of despair and futility."[15] Heisenberg had maintained for years that a uranium-based bomb would have to be too massive and too large to be practically carried by a plane. Germany would need too much enriched uranium, and even a more resource-rich country like America would need decades to process it. So when they digested the news of August 6, reactions spanned Heisenberg's disputing the report to Walther Gerlach's threat of suicide given the now-apparent failure of German science. A horrified Hahn said he was glad Germany hadn't built such a bomb, and he wondered if, "all uranium should be sunk to the bottom of the ocean."[16] A younger physicist suggested they never truly *wanted* to build a uranium weapon. (In fact, this became a rallying point for the German physicists: We invented fission, but we were never so barbaric as to weaponize it.)

Goudsmit recalled one different voice among the Germans. "Von Laue was probably the only man among the ten interned on that English estate who realized thoroughly the world-shaking effects this Hiroshima bomb would produce everywhere, in ever-widening circles."[17] In this, von Laue's prescience matched that of his mentor, Max Planck.

As Lise Meitner's close friend, and as the long-time leader of German science, Planck had long understood the energies available from nuclear fission. The industrial scientist and inventor Manfred von Ardenne, who contributed to Germany's ill-fated nuclear project, once recalled spending a wartime winter day with Planck. It was 1940 and Planck, fully retired, had enjoyed a tour of von Ardenne's laboratory. Planck accepted a ride home to Wangenheimstrasse, but he provoked his host a bit on the way. He snagged a Nazi pamphlet from von Ardenne's backseat and read headlines portending coming attacks against England. "So you think America will declare war on us one day?" Planck asked, almost rhetorically, and von Ardenne said he did. "Then we are expecting the same thing," Planck continued. "The current military successes should not deceive a scientist who has learned to think critically. Unfortunately, a lot of people are still deceiving themselves. ... I am very worried."

While driving, von Ardenne regarded the old physicist and asked if he was talking about nuclear fission, about the possible consequences.

"The consequences will be unimaginable," Planck said. "If this instrument of power gets into the wrong hands. . . ."

His companion was less comfortable with such a dangerous conversation in these times, especially with the implication of "wrong hands," and he emphasized civilian uses. "It's nature's most powerful source of energy," von Ardenne said.

"Yes," replied a tired Planck. "And it must be used to the benefit of mankind . . . but it will happen differently."[18]

The news of Hiroshima and Nagasaki only confirmed Planck's unusual bout of pessimism. In a postwar letter to a friend, Planck suggested that if the Russians had developed the bomb before the United States, it would have meant, "the end of our culture" for Germans.[19]

The afterglows of Hiroshima and Nagasaki create another striking bookend for the times of Max Planck. Just 75 years earlier, when his older brother Hermann died in the Franco-Prussian war, breech-loaded rifles were proving themselves superior to muzzle-loaded firearms, and Germany's new steel cannons were among the world's most sought-after weapons. Planck was born into an 1858 that employed candlelight and horse-drawn carriages, but he came to hear radio-transmitted symphonies and watch airplanes crisscross the sky. He joined a physics in the nineteenth century that was reportedly nearing completion. There was as yet no Planck's constant to stand between a scientist and the pursuit of exact knowledge. The atom was unlikely, light a pure wave, and energy indivisible. Time and space stood absolute and inflexible before our rulers and clocks. Physics was a far flung outpost of science then, with a tiny clan. Germans jokingly confused it with forestry. But as Max Planck reached his last years, physicists stood like shamans for a nuclear age. Their ideas flashed and echoed around the globe.

Coda: 1945–1947

In August of 1945, Planck had recovered well enough to leave the hospital. He and Marga established themselves as best they could in Göttingen, where Max had family. Planck's brother Adalbert had died in Göttingen just before the war, but Adalbert's daughter Hildegard "Hilde" Seidel lived there with her husband and daughter. Their home at Merkelstrasse 12 became the Plancks' last dwelling.[1] According to Marga's journal entries, Hilde gladly received them and gave them the main bedroom.[2] Although they were fortunate to have shelter, the postwar months were still incredibly challenging. As the industrialist Carl Still (once the Planck's Rogätz host), noted with the approach of winter: "The heat is more important than the food."[3]

In his last years, Max Planck had returned to family headwaters. His great-grandfather, Gottlieb Jakob Planck, had spent his career there as a theology professor.[4] If the Plancks could find an additional silver lining, their new environs were only lightly damaged compared with most German cities. (To this day, Göttingen retains a startling range of prewar architecture that has been lost elsewhere.)

We see glimpses of these last years in Marga's correspondence. In a letter to Hilde's sister in America, she fretted about the imposition she and Max placed on their hosts, "but I don't how I would now organize my own household without all the furniture, linen and necessary things which daily life requires." She thanked Frieda for the care packages, saying they had helped restore Max, even though the "little ones" (Planck's great-grandchildren), tore them open and ransacked them first. "Uncle Max is really feeling the improved nutrition," she wrote. "Now his memory is suffering horribly, and that is very sobering for me, but thank God he doesn't notice. Along with that he is rarely not being stubborn—he is after all a 'Plank.' He can barely get over the loss of Erwin." She also mentions her own son, who remained in the Russian zone. "Hermann struggles . . . and goes it alone in Berlin, as all others who don't follow communism must."[5]

After World War II, the German physics community would eventually start to reconvene in Göttingen. Hahn and Heisenberg, for instance, set up their postwar careers there. But while the 10 were still interned at Farm Hall, the Allies turned to the 87-year-old Planck to again serve his homeland, and he returned in 1945 to direct the Kaiser Wilhelm Society (KWG). As he wrote at the time, "I am not one of those who let themselves be bitter."[6]

Released from Farm Hall, Max von Laue visited his old friend and mentor, though it distressed him. "He has grown slumped and old, complaining of arthritis and bad pain," he wrote to Lise Meitner in the spring of 1946. "He inquired about physics, but he does not talk about what he went through." He reassured Lise that Planck still played piano for at least 15 minutes every night, even though two fingers of his left hand were too stiff to use (Figure C.1).[7]

In the summer of 1946, the Royal Society of London held a delayed celebration for Isaac Newton's 300th birthday. The war had eclipsed any notion

Figure C.1. Max Planck during his daily piano time in Göttingen, 1946.
Courtesy Archiv der Max-Planck-Gesellschaft, Berlin-Dahlem.

of celebrating the actual anniversary, Christmas of 1942. They invited one and only one German scientist to the festivities. The Royal Society had long appreciated Max Planck, having elected him a Foreign Member in 1926 and then awarding him their highest honor, the Copley Medal, in 1929.[8] Now, in the postwar festivities, he must have winced when hearing his official introduction, "Max Planck from no country." What had been Germany was now occupied by four different nations. After stirring and sustained applause for Planck, the president of the Royal Society issued a correction: "The announcement should have been, Professor Max Planck from the World of Science."[9] The press noted the importance of Planck's attendance and recognized his personal suffering at the hands of the Nazis. The *New York Times* headline read, "Scientists honor Isaac Newton in London; Planck Brought from Berlin for Ceremony," and the short article, referring to him as a "shriveled figure," noted, "His son was executed following the attempt on Hitler's life in 1944."[10]

Lise Meitner was on hand and enjoyed her first postwar visit with her former boss. She later wrote to Marga that seeing her former teacher was, "like a gift from heaven . . . the purity and integrity of his personality have resisted all the years."[11] But she wrote to a friend about new qualities, previously inconceivable in Planck: physical weakness and a short-circuiting memory.[12]

Otto Hahn also noted Planck's frailty. "I was somewhat shocked by the rapid decline in Planck's intellectual strength," he wrote later to Meitner. "He has aged greatly in the last six months, when I compare him, for instance, to the King of Sweden."[13] This last detail shows not only a blunt insensitivity to Planck but near cruelty to Meitner. Just three months after the first use of atomic weapons, the prize committee in Stockholm awarded Otto Hahn its Nobel Prize for Chemistry. Although the official citation mentioned a "dangerous" side to the technology, it emphasized the long-term prospects for peaceful use. And while it also mentioned Lise Meitner as a contributor, she shared neither the prize nor the recognition of her long-time collaborator.[14]

Tensions between Hahn and Meitner lingered for years, but this owed much more to the topic of Nazi Germany than any scientific snub. She felt Hahn compartmentalized Nazi atrocities as separate from Germany itself. Even von Laue disturbed her in this way. Although he'd been one of the most public German scientists in opposing Nazism, he fought to create distance between German citizens (particularly German scientists) and their Nazi government. After the war, von Laue wrote an article called "The Wartime Activities of German Scientists." This offering stumbled onto a brightly lit public stage: the aftermath of Goudsmit's public accounts of the Alsos Mission and his revelations that German physicists had indeed worked for the Nazis on a secret nuclear project. Von Laue wrote "a few words of protest" against international claims that German scientists worked with the Nazis and even exploited concentration camp labor.[15]

Meitner wrote to Hahn about von Laue's article.

> Is it really justifiable to say that the majority of scientists were against Hitler from the beginning? . . . When Planck held the memorial for Haber, Laue and Huebner were the only professors to dare to come to it: At the same time the Chemical Society and the Glass Engineering Society . . . had forbidden their members from attending. . . . Doesn't this all seem to show that the subordination to the Hitler ideas was very prevalent and that the opposition . . . was a minority? I truly do not have the intention of saying unpleasant things with these observations, but I am afraid that with his inclination to defend everything that has happened—out of understandable attachment to Germany—Laue is not helping Germany but risks achieving the opposite.[16]

She was similarly dismayed when her old colleagues defended the name of Kaiser Wilhelm. The Allies had come to see the Nazis as part of a larger bellicose pattern in Germany. Hyperbolic Nazis aside, they worried about the warrior mystique surrounding the name of the one-time Emperor, he of pointy helmet and military regalia. They didn't want a rebuilt Germany to look with pride on the Kaiser's name, so they forced a sort of rebranding. The Allied command started dissolving the KWG sector by sector, in 1946. Planck, Hahn, von Laue, and seven other Nobel Prize–winning German scientists telegrammed their protest to a governing American general, asking that he reconsider. They sought to draw a line between Nazis and the Society's independent work. General Lucius Clay replied politely, denying the request, but assuring them that the future of their scientific institutes were "being carefully considered" and that new policies would soon be announced.

Meitner wrote to Hahn with patient logic regarding the KWG.

> What the best people among the English and the Americans wish is that the best Germans understand that this unfortunate tradition, which has brought the whole world and Germany itself the greatest misfortune, must finally be broken. And a small token of this insight is to change the name of the KWG.

What meaning is there in a name, when one is concerned with the existence of Germany, and with it, Europe?[17]

The old title, and the image of the empire's monarch, quietly faded, as one uniting name rose to replace it. The Max Planck Society for the Advancement of Science was first founded in the British Zone in 1946.[18] Having never embraced the Nazi party and having lost a son to the Führer's vengeance, Max Planck had a unique standing among Allies and Germans alike. Here, nearing the end of his turbulent life, the widespread comfort and respect attached to his name reminds one of his early reports from grade school, where he was well liked by teachers and classmates alike. Planck gave his blessing to the use of his name, as an aid to reconstruction. He said he especially wanted to preserve a society committed, "only to the truth of science, independent of all currents of a particular time."[19] The rebranding campaign spread to the American and French zones, and the official launch year for the Max Planck Society—today boasting 82 separate institutes and thousands of employees across Germany—is now recognized as 1948.

Soviet-controlled East Germany claimed Planck as well. They underlined that Planck, just like Lenin, had verbally fought against the philosophy of Ernst Mach. During the 1958 "pan German national celebration" of his 100th birthday, the nation's central committee announced, "only the working class that built socialism and defends world peace has the right to celebrate the great scientist Max Planck. . . . What Planck and a younger generation of physicists has created, capitalism can no longer contain."[20]

The recovering nations of Europe faced an unusually nasty winter in late 1946 and early 1947. The continent teetered with millions of homeless refugees, demolished infrastructure, fuel shortages, and scarce nourishment. Even mild cold would have been difficult, but Europe endured one of the worst winters on record. "But I can assure you of one thing," Marga wrote to Lise Meitner, "without the help of our friends abroad, my husband would very certainly not have survived the winter."[21] Care packages kept many Germans alive. Despite incredible frailty and the bleak conditions, Planck delivered his last guest lecture in late March 1947, just a month shy of his eighty-ninth birthday.[22]

Planck's former student Gabriele Rabel provides a window into his last days, as she corresponded with Marga from London. After Rabel learned of the Plancks' Göttingen address, she sent along one of her new research papers with a short note.

Marga replied,

> Unfortunately, he can no longer write himself. After overcoming surprisingly well a severe pneumonia last winter, he had a fall six weeks ago. . . . I am wondering whether he can at all recover. Alas, he has a very sad old age. Loss of his home, loss of four children, ill and feeble—what good are now to him all his honours?

"I hastened to send a few of the dainties which we enjoyed here and which were unknown in starving Germany," recalled Rabel. "Marmalade and tea and so forth."

Marga thanked Rabel, but Max

> could not enjoy the treasures which you had meant for him. On the fourth of October, he has closed his eyes for ever. My husband might well have recovered from the consequences of his fall, but haemorrhages in the brain were added, and now no medical art could help . . . He was so tired, so exhausted. . . . I must not grudge him his rest and I look gratefully back on the life I was allowed to live at the side of this rare man.[23]

At Planck's memorial, the Lutheran theologian Friedrich Gogarten presided, weaving bits of Planck's own writings seamlessly into his comments.[24] Max von Laue then delivered an address to those assembled in Göttingen's Albani Church. As his friend and mentor would have liked, he focused primarily on Planck's work, and especially the breakthrough of 1900. In concluding his remarks, von Laue gestured to the many wreaths sent by museums, academies and scientific societies, and spoke to simplicity, one of Planck's favorite qualities.

"And here is a plainer wreath, without any streamers. It was placed here by me on behalf of all his pupils, among whom I count myself, as a perishable token of our never-ending affection and gratitude."[25]

Max Planck was buried in the City Cemetery in Göttingen, where his memorial is joined now by those of Max von Laue and Otto Hahn, among many others.

After the funeral, Marga received a letter from Princeton, New Jersey, of familiar script and unmistakable prose.

Albert Einstein wrote:

> Now your husband has finished his days after he achieved greatness and experienced much bitterness. His gaze was fixed on the eternal things, and yet he took an active part in all that was human and he lived in the temporal sphere. How different and better the human world would be if there were more such

unique people among the leaders. So it seems not to be, as the noble characters in every time and every place must remain isolated without being able to influence the events around them.

The hours that I spent in your home, and the many conversations that I conducted in private with the wonderful man will for the rest of my life belong to my beautiful memories. It cannot change the fact that a tragic event tore us apart.

In today's loneliness, may you find comfort in that you have brought sun and harmony into the life of this revered man. From a distance, I share with you the pain of parting.[26]

Despite Planck's advising Einstein to avoid dreaming of outer space, the satellite bearing Planck's name now glides in Earth orbit, looking where the scientist himself never did. As of this writing, the Planck mission has successfully scanned the entire background signal of the universe: the hiss we hear in the cosmic speakers after the music stops. Although the radiation perfectly fits Planck's radiation law, the mission has also found a few surprises. The primordial fireball was not symmetrical—it appears to feature an unexpected cool patch, as if the infant universe brushed a cheek against a cold window. In a methodical attempt to explore one question, the satellite has uncovered a different and even more puzzling question. One has to believe its namesake would nod with approval. As he wrote shortly before Hitler's rise and Einstein's departure:

For it is just this striving forward that brings us to the fruits which are always falling into our hands and which are the unfailing sign that we are on the right road and that we are ever and ever drawing nearer to our journey's end. But that journey's end will never be reached, because it is always the still far thing that glimmers in the distance and is unattainable. It is not the possession of truth, but the success which attends the seeking after it, that enriches the seeker and brings happiness to him.[27]

Appendix

A Modern Look at the Thermal Radiation Spectrum

The main narrative chronicles Planck's work on thermal radiation, including his breakthrough of 1900, but it stays with Planck and the limited tools available to him. Here I want to provide a more contemporary understanding of the underlying physics. Even though we'll wade a bit further into technical waters, I still write this with a general audience in mind, including a hopefully appetizing analogy.

In 1900, Planck didn't have the benefit of understanding the true nature of light and all electromagnetic radiation. We now know it is composed of massless bits of energy called "photons."

It turns out that the beautiful curve with its narrow peak describes a sort of tug-of-war between fundamental influences—the peak is just a marker for the central knot in the tug-of-war rope. (The curve appears in Figure P.1, for the cosmic microwave background radiation, and in Figure 5.4, for the black bodies examined in the German laboratories of the 1890s.) On one end, we have to figure out all the different ways a photon could exist, given a certain energy. This is roughly like holding a doll, figuring out which joints bend, and then finding all the different poses the doll can hold. The greater the photon energy, the more imaginable ways (or modes) we can imagine for the appearance of a photon. On the other end of the rope, we have a reality check, which is the actual number of available photons reporting for duty at a given temperature. Each of these effects would make a very simple graph on its own, just sloping downward to the left or to the right respectively, but put together, their opposite tendencies result in a "tie," evidenced by a central peak showing the most popular color of black-body photons at a given temperature.

Our analogy has us sitting and watching a bakery from the outside—that is our black-body experiment. We can't see inside, but we can observe (and eat) the various types of baked goods that emanate from within it. We want to model exactly the *spectrum* of goodies produced by the bakery each day—that is the weight of baked goods produced at each specific pastry size, from tiny donut holes to enormous wedding cakes. In this way, we aim to sketch a curve that would show the weight of baked goods per day on the vertical (y) axis versus the individual sizes of different goodies on the horizontal (x) axis. There are three factors that are crucial to figuring out the resulting curve, and if we multiply these three together, we should be able to get something close to the desired curve—literally a *radiated bakery spectrum.*

1. The individual weight for each pastry size.
2. The total options of possible baked goods that can be made at a certain weight. For example, starting with the smallest item, a donut hole, the baker has two basic options: plain or chocolate. Once you reach the size of a cinnamon roll, you could also have various muffins of the same size. *In general*, the baker has more and more options as he considers items of greater size. Think of the nearly infinite ways one can scaffold and decorate a wedding cake, versus just the two flavors of donut holes.
3. The internal decision of the bakery concerning how to allocate its limited resources. A bakery could send out infinite cakes and cream puffs if it had infinite ingredients, but the basic weight of ingredients is limited, assuming a fixed budget. The bakery has some kind of *internal logic* it uses, typically making just a few big cakes and a lot of muffins and rolls, because there is greater demand for items customers can take in hand to a boring morning meeting (or physics class).

Factors 1#1 and 2 increase the output weight of the bakery products as you increase the pastry size, so they would suggest a curve or line with an upward slope on our proposed graph. However, the reality check comes from factor 3. There is a limited amount of dough, and the bakery can't devote itself to just 600-lb super-cakes. It typically makes more of the smaller items and fewer of the larger items. So if we just plotted factor 3, we would get a curve with a downward slope (Figure A.1).

When we multiply these three factors together, the net effect is to make a hump in the middle (see Figure A.1b). At the low end, the bakery makes dozens of donut holes, but each one is lightweight, so a relatively small collective weight of donut holes rolls out of the bakery every day. On the high end, there are very few enormous wedding cakes produced on the average day, so the collective weight of enormous items is also pretty small. The greatest total output weight of baked goods comes in the form of handheld items like muffins, rolls, and scones. We have lots of different kinds, with a decent heft per item.

This analogy establishes our principal players well enough. The radiation spectrum of a black body shows radiated power on the vertical axis, plotted against frequency on the horizontal axis, and we will multiply three different effects together in order to achieve that full spectrum. The chunks of radiation (which we now call *photons*) are analogous to pastries.

The energy of each photon is analogous to the weight of each pastry, and the frequency of each photon is analogous to generic pastry size. So the first important

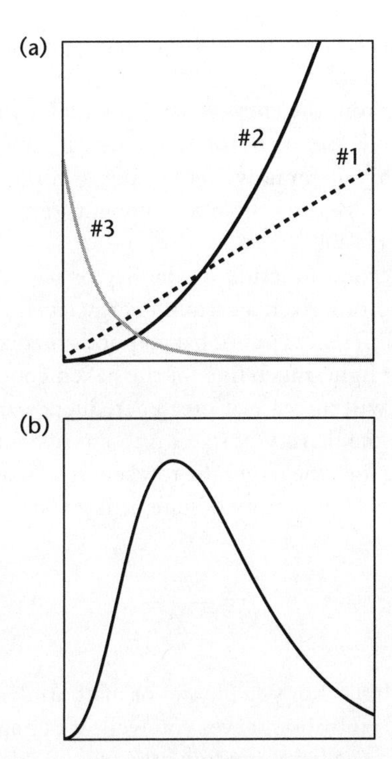

Figure A.1. These plots represent the sizes of individual pastries on the horizontal axis and the subtotal of the bakery's weight output at a given pastry size on the vertical axis. Plot (a) shows each bakery factor on its own, labeled #1, #2, and #3. Plot (b) shows the result of multiplying all those factors together: the *total bakery output spectrum*, in terms of weight output at each pastry size. (Note: The horizontal range of pastry sizes are consistent from graph a to b, but the vertical scale, total pastry weight, is much larger in b than a.)

function is one of Planck's most insightful and enduring suggestions, that the energy of radiation from a black body must be related directly to its frequency, or

$$E = hf \tag{1}$$

where E is the photon's energy, h is Planck's universal constant, and f is the frequency of the electromagnetic radiation. This equation dates to Planck's December 1900 presentation, where he claimed that "the most essential point of the whole calculation," was that energy would be emitted in chunks of energy hf. This was *the* conception moment for quantum theory, with a granular treatment for energy.[1] We finally figured out that the best model of a bakery's behavior has it selling *units* of pastries and not just squeezing raw dough into the hands of customers.

(An important note from the chef: Where this analogy leaves a messy kitchen is in the analogy of pastry size to photon frequency. It's not a clean conceptual fit, but it's the best I can whip up for now. In the bakery world, it seems obvious that a pastry's weight tracks its size in a simple way, but equation 1, in the world of physics, was much more surprising.)

Physicists call the second function a "density of states," and it describes the different options of photons given a certain energy level. In our bakery analogy, this is the number of different types of baked goods that could possibly be made at a certain weight. For light, much like for the baked goods, the number of possible types of photons will increase at greater frequency and energy levels. One way of thinking about the density of states is to consider an apartment building. The larger the building (i.e., the larger the total energy "space"), the more photon housing options we have. The density of states ρ, is given by

$$\rho = \frac{8\pi f^2}{c^3} \tag{2}$$

where c is the speed of light. For you physics or math students, if the 8π looks like a geometric factor, your intuition serves you well. If our apartment building was spherical, and everyone needed a window unit, the total possible number of different tenants would be proportional to $4\pi f^2$, the surface area of the apartment complex. We end up with an extra factor of two for photons because of their so-called polarizations. The photon still has wave-like qualities, and as it moves toward you, it can wiggle up-and-down or side-to-side, and these two options, like vanilla or chocolate, increase the total number of photon options for any frequency, so that's why the density of states has the factor of $8\pi f^2$.

Before we move to the third factor, let's look at the first two. Based on just the energy per photon and the total photon options at any given frequency, we see both functions just increasing. So given these two, we would expect a black body to radiate more and more photons for an ever-higher frequency of light. In fact, when the classical picture of physics was better refined, by 1911, the young physicist Paul Ehrenfest labeled this expectation the "ultraviolet catastrophe." Ultraviolet light has higher frequencies than visible light, and classical physics was predicting that even your clothes, at room temperature, would glow a dull blue color to your eyes and then send out enough ultraviolet to give you their own brand of sunburn. Such a catastrophic mismatch between reality and classical theory underlined the need for new thinking, helping motivate the first Solvay conference on quantum theory.

So now we move to the third function, the distribution of photons given a naturally limited amount of available energy. Everything to this point has assumed an infinite supply of available energy, but now we bring in some reality. In our analogy, this is the mysterious way in which a bakery decides to allocate its supply of ingredients. The ambient temperature, much like a supply of bakery ingredients, provides the energy needed to cook up some photons.

Planck intuited this distribution, nature's own choice of pastries, in his formula of 1900, but it was not well understood for many years. Hendrik Lorentz took an early run at understanding the distribution by 1910, and we must acknowledge the wholly original and now time-tested approach of Albert Einstein in 1917, when he built a new and surprisingly simple theory for analyzing the emission and absorption of radiation.[2] In the 1920s, the young Indian physicist Satyendra Nath Bose, in an exchange of letters with Albert Einstein, came to propose the precise distribution function for photons (and similar fundamental critters). So the third factor, buried within Planck's 1900 formula, is today called the Bose-Einstein distribution.

$$g(f) = \frac{1}{e^{\left(hf/kT\right)} - 1} \tag{3}$$

The k here is Boltzmann's constant, T is temperature, and f is frequency. Note the important ratio hf/kT; this is simply comparing, by analogy, the weight of one pastry to the total weight of available ingredients, and it determines the whole shape of the resulting function, g. We could write a whole book about the remarkable implications of this function—it contains the magic of thermal radiation and its universal character. Unlike human bakeries, which all make their own unique decisions of how to allocate their dough, every object in the universe obeys the same rules for distributing thermal energy to photons. Though some people bemoan restaurant chains, all of nature's black-body bakeries practice a hyper-uniformity. And *this* is precisely why the thermal radiation curves are so universal, and why they are only altered by temperature (i.e., just the amount of dough available).[3]

Unlike the first two equations, this third equation *decreases* dramatically as frequency (and therefore energy) increases. The bakery figure's curve #3 actually shows the correct function. And so the competing effects join in battle. Multiplying all three together, we obtain the thermal radiation curve, as set out perfectly by Max Planck in 1900.

$$(1) \times (2) \times (3) = \left(\frac{8\pi hf^3}{c^3}\right)\left(\frac{1}{e^{hf/kT} - 1}\right) \tag{4}$$

Planck's first derivation for this was flawed. Because our approximate derivation here benefits from hindsight, it uses ideas and tools (like equation 3) that Planck did not have in his available kit at the time. These tools were also largely unavailable when he penned his so-called second quantum theory in 1912 and 1913.[4]

While he failed to perfect a derivation from points A to Z, he birthed critically correct and wholly original steps along the path in each attempt, and these steps persist into the present. Equation 1 was a primary result of his first attempted derivation. Just as amazing, but less well appreciated, is a result of his second attempted derivation. In Planck's second quantum theory, his modified statistical approach created a small artifact: No matter the temperature of an object, and no matter the color of light one examined, there would always be an extra little "chunklet" equal to *half* of one quantum of energy. It was a trifle and could largely be ignored in most cases. Far from a mistaken slip of Planck's pen, the effect, the "zero-point energy," is real. It describes a sort of very low-level but ubiquitous seething of energy in the fabric of the cosmos itself. It was an unavoidable result of Planck's mathematical approach, and he even apologized for introducing such an ugly wart to quantum theory. He wrote a *mea culpa* to Paul Ehrenfest in 1915.

> I fear that your hatred of the zero-point energy extends to [my theory overall]. But what's to be done? For my part, I hate discontinuity of energy even more than discontinuity of emission. Warm greetings to you and your wife.[5]

The zero-point energy lingers in high-level theoretical conversations 100 years later. Astrophysicists are plagued by the obvious but unexplained rapid expansion of the universe. Far from a residual effect of a Big Bang, our universe appears to be expanding faster and faster, like an explosion that is then powered by more and more explosions. In looking to explain the mysterious "dark energy" causing this accelerated expansion, physicists often start with two historical factors, emanating respectively from Einstein and Planck.

The first is Einstein's "cosmological constant," his proposal for propping up the universe against gravity. He sought a way to keep everything from collapsing into one enormous clump. The other is Planck's zero-point energy that fills our quantum reality, giving the great void a "vacuum energy," always on hand. I refer interested readers to Helge Kragh's very readable account of how the Berlin chemist Walther Nernst first interjected Planck's zero-point energy into cosmology just a few year's after hearing Planck unveil the second quantum theory.[6] Either factor could, hypothetically, help explain an outward pressure in the universe, but as of this writing, neither suspect can be blamed for the so-called dark energy we observe.

Notes

PREFACE

1. R. Wilson, "Discovery of the Cosmic Background Radiation." *George Gamow Symposium; ASP Conference Series*, 129 (1997): 84–94.
2. Original COBE satellite data using FIRAS instruments, made publicly available and adapted on Wikimedia commons.
3. A. Einstein. AAAS bulletin, April 14, 1948. American Philosophical Society archives.

CHAPTER 1

1. K. Hentschel, *Physics and National Socialism*, Exhibit 55, "'White Jews' in Science," *Das Schwarz Korps*, July 15, 1935 edition.
2. J. Heilbron, *The Dilemmas of an Upright Man*, pp. 210–214.
3. A. Pufendorf, *Die Plancks*, p. 463. Erwin's friend Georg Thomas later wrote that he saw Erwin while they were both in custody, and he relayed that Erwin had been tortured for weeks.
4. A. Pufendorf, *Die Plancks*, p. 458.
5. F. v. Schlabrendorff, *The Secret War Against Hitler*, p. 295.
6. Plötzensee Memorial Center Hüttingpfad, D-13627, Berlin.
7. M. Planck to M. v. Laue, January 29, 1945, as in J. Heilbron, *Dilemmas*, p. 195.
8. Plötzensee Memorial Center Hüttingpfad, D-13627, Berlin.
9. See, for instance E. W. Kolb and M. S. Turner, *The Early Universe*.
10. A. Hermann, *Planck*, p. 24.
11. P. Rife, *Lise Meitner and the Dawn of the Nuclear Age*, pp. 32–33.
12. A. v. Pufendorf, *Die Plancks*, p. 451; K. Hentschel, *Physics and National Socialism*, Exhibit 64, July 21, 1938 letter from H. Himmler to W. Heisenberg, pledging political support because "you were recommended by my family."

CHAPTER 2

1. J. Heilbron, *Dilemmas*, p. 192.
2. A. v. Pufendorf, *Die Plancks*, p. 439.
3. Ibid., p. 438.

4. Ibid., p. 431.
5. J. Heilbron, *Dilemmas*, p. 98.
6. Dr. G. Rabel, as recounted to T. S. Kuhn, 1963. American Institute of Physics Archives.
7. J. Heilbron, *Dilemmas*, p. 2.
8. M. Planck to Emma Lenz. November 21, 1917. Archiv der Max-Planck-Gesellschaft, Dept. Va, 11 Rep, No. 757, Max Planck collection.
9. M. Cardona and W. Marx, "Max Planck—A Conservative Revolutionary." *Percorsi*, 24 (2008): 39; also see C. Seidler, "Namens-Überraschung: Gestatten, Marx Planck," *Spiegel Online Wissenschaft*, April 24, 2008. Max would be a shortening of Maximilian, whereas Marx, if not written in error, could have been an outmoded translation of Marcus.
10. J. Heilbron, *Dilemmas*, p. 1.
11. *Deutsches Geschlechterbuch Band 178*, "Family peiper," p. 183.
12. T. Mann, *Buddenbrooks*, p. 51.
13. Ibid., p. 42
14. A. Hermann, *Planck*, p. 8.
15. D. Hoffmann, *Max Planck: Die Entstehung der modernen Physik*, pp. 10–11.
16. J. F. Cooke, *Great Pianists on Piano Playing*, pp. 241–245.
17. A. Hermann, *Planck*, p. 25. Taken from Max von Laue's comments in 1958, at a memorial to commemorate Max Planck's 100th birthday.
18. *Deutsches Geschlechterbuch Band 178*, "Family peiper," p. 183.
19. J. Heilbron, *Dilemmas*, p. 3.
20. M. Planck, *Scientific Autobiography and Other Papers*, p. 14.
21. Ibid.
22. J. Heilbron, *Dilemmas*, p. 10.
23. M. Planck, *Scientific Autobiography and Other Papers*, p. 19.

CHAPTER 3

1. After an interruption at the end of World War I, Planck was awarded the 1918 prize, and Stark the 1919 prize, in 1920.
2. As in W. Isaacson, *Einstein*, p. 289.
3. J. Heilbron, *Dilemmas*, pp. 113–114.
4. L. Meitner to E. Schiemann, June 30, 1943, as in A. v. Pufendorf, *Die Plancks*, pp. 433–434.
5. L. Meiter to M. v. Laue, May 30, 1943, as in A. v. Pufendorf, *Die Plancks*. pp. 433–434.
6. R. L. Sime, *Lise Meitner: A Life in Physics*, 1996; P. Rife, *Lise Meitner and the Dawn of the Nuclear Age*, 1999.
7. P. Rife, *Lise Meitner and the Dawn of the Nuclear Age*, pp. 3–11.
8. R. L. Sime, *Lise Meitner*, p. 13.
9. Ibid., p. 15.

10. Ibid., p. 25.
11. M. Planck, C. Runge, et al., *Brieftagebuch*, pp. 173–174, entry of March 28, 1909.
12. As in P. Rife, *Lise Meitner and the Dawn of the Nuclear Age*, p. 32.
13. As in R. L. Sime, *Lise Meitner*, p. 37.
14. As in P. Rife, *Lise Meitner and the Dawn of the Nuclear Age*, p. 25.
15. R. L. Sime, *Lise Meitner*, p. 36.
16. P. Rife, *Lise Meitner and the Dawn of the Nuclear Age*, p. 30.
17. R. L. Sime, *Lise Meitner*, p. 35.
18. R. L. Sime, *Lise Meitner*, pp. 109–110; P. Rife, *Lise Meitner and the Dawn of the Nuclear Age*, p. 87.
19. P. Rife, *Lise Meitner and the Dawn of the Nuclear Age*, pp. 25, 33.
20. R. L. Sime, *Lise Meitner*, p. 204.
21. Ibid.
22. R. L. Sime, *Lise Meitner*, p. 209.
23. Ibid., pp. 306, 336.
24. Deportation Memorial, Grunewald Station, Deutsche Bahn AG.
25. E. Lichtblau. "The Holocaust just got more shocking." *New York Times*, March 1, 2013.
26. K. Hentschel, *Physics and National Socialism*, Exhibit 108, L. Meitner to O. Hahn, June 27, 1945.
27. R. L. Sime, *Lise Meitner*, p. 306.
28. As in J. Heilbron, *Dilemmas*, p. 207.
29. As in R. L. Sime, *Lise Meitner*, p. 37.
30. O. Frisch and J. A. Wheeler, "The Discovery of Fission." *Physics Today*, 20 (1967): p. 47.
31. P. Rife, *Lise Meitner and the Dawn of the Nuclear Age*, pp. 106–107.
32. As in R. L. Sime, *Lise Meitner*, pp. 333–334.
33. T. Powers, *Heisenberg's War*, p. 207.
34. Ibid., p. 283.

CHAPTER 4

1. Royal Air Force digital archives.
2. W. G. Sebald, *On the Natural History of Destruction*, p. 19.
3. J. Heilbron, *Quantum Mechanics*, p. 34.
4. M. Planck, Nobel Prize lecture, 1920.
5. As in J. Heilbron, *Dilemmas*, p. 40.
6. Ibid., p. 61.
7. As in J. Heilbron, *Dilemmas*, pp. 41–43.
8. G. Rabel interviewed by T. Kuhn in 1963. American Institute of Physics Archives.
9. M. Planck, *Scientific Autobiography and Other Papers*.

10. M. Planck, "Religion and Natural Science," in *Scientific Autobiography and Other Papers*.
11. M. Planck to W. H. Kick, June 18, 1947, as in J. Heilbron, *Dilemmas*, p. 198.
12. V. Viereck, interview with A. Einstein, *Saturday Evening Post*, October 26, 1929.
13. J. Heilbron, *Dilemmas*, p. 191.
14. A. J. P. Taylor, as in W. G. Sebald, *On the Natural History of Destruction*, p. 18.
15. W. G. Sebald, *On the Natural History of Destruction*, p. 17.
16. A. Speer, *Erinnerungen*, as in W. G. Sebald, *On the Natural History of Destruction*, p. 17; Sebald, pp. 3–4.
17. G. A. Craig, *Germany 1866–1945*, p. 760.
18. Ibid., p. 761.
19. P. Englund, *The Beauty and the Sorrow*, p. 288.
20. M. Planck to E. Planck, October 24, 1943, as in A. Pufendorf, *Die Plancks*, p. 436.
21. J. Heilbron, *Dilemmas*, p. 192.
22. C. Rhoads, *Wall Street Journal*, February 26, 2003; BBC News, 22 October, 2003. 60th anniversary of Kassel Bombing.
23. M. Planck to E. Planck, October 24, 1943, as in A. v. Pufendorf, *Die Plancks*, p. 437.
24. W. G. Sebald, *On the Natural History of Destruction*, p. 30; Nossack, as in W. G. Sebald, *On the Natural History of Destruction*, p. 41.
25. W. G. Sebald, *On the Natural History of Destruction*, pp. 85, 83.
26. M. Planck to E. Planck, October 24, 1943, as in A. Pufendorf, *Die Plancks*, p. 437.

CHAPTER 5

1. A. von Pufendorf, *Die Plancks*, p. 434.
2. Ibid., p. 441.
3. As in W. G. Sebald, *On the Natural History of Destruction*, p. 65.
4. M. Planck, *Über den zweiten Hauptsatz der mechanischen Wärmetheorie*, 1879, as in T. Kuhn, *Black-Body Theory and the Quantum Discontinuity 1894–1912*, p. 15.
5. M. Planck, *Scientific Autobiography and Other Papers*, p. 19.
6. Ibid.
7. R. L. Sime, *Lise Meitner*, p. 13.
8. M. Planck, *Scientific Autobiography and Other Papers*, p. 19.
9. Ibid., p. 20.
10. M. Planck, C. Runge, et al., *Brieftagebuch*, p. 90, entry of October 7, 1883.
11. C. Jungnickel, *Intellectual Mastery of Nature*, p. 29.
12. A somewhat common story. See, for instance, A. Norton et al. (ed.), *Dynamic Fields and Waves*. London: Institute of Physics Publishing, 2000, p. 83.
13. E. Mach in *Physikalische Zeitschrift*, 1910, as in J. Heilbron, *Dilemmas*, p. 54.

14. M. Planck to M. v. Laue. August 5, 1910, as in J. Heilbron, *Dilemmas*, p. 55.
15. M. Planck, *Physikalische Zeitschrift*, 11 (1910): 1187–1188, as in J. Heilbron, *Dilemmas*, p. 55.
16. A. v. Pufendorf, *Die Plancks*, p. 156.
17. J. Heilbron, *Dilemmas*, pp. 123–124.
18. M. Planck to A. Einstein, October 26, 1918. *The Collected Papers of Albert Einstein.*
19. M. Planck, Nobel Prize Lecture, 1920, as in J. Murphy's preface to M. Planck's *Where Is Science Going?*
20. M. Planck, *Scientific Autobiography and Other Papers*, p. 153
21. For the most physics-interested audiences: although frequency and wavelength are simply inversely related, the actual location of the peak in these plots varies between the different representations because of specific mathematical details related to the choices made in measuring the spectra. As physicist Mark Heald cautions, one must keep track of the "bookkeeping" involved. Please see his article, "Where Is the Wien Peak?" *American Journal of Physics*, 71 (2003): 1322. And for a discussion of how this difference in the peak relates to arguments of evolution of human vision (in which our sensitivity to light does not quite match the peak radiation from the sun), see Soffer and Lynch's article, "Some Paradoxes, Errors, and Resolutions Concerning the Spectral Optimization of Human Vision." *American Journal of Physics*, 67 (1999): 946–953.
22. M. Planck, *Scientific Autobiography and Other Papers*, p. 24.
23. As in J. Heilbron, *Dilemmas*, p. 7.
24. D. Hoffmann, et al. *Quantum Theory Centenary*, pp. 19–30.
25. Ibid., pp. 39–40.
26. My biologist wife would like to point out that these three would never occupy the same forest, but that is hardly the current point of the analogy.
27. M. Planck, *Scientific Autobiography and Other Papers*, pp. 34–35.
28. As in J. Heilbron, *Dilemmas*, p. 25.
29. Ibid., p. 190.

CHAPTER 6

1. G. Pihl, *Germany: The Last Phase*, pp. 228–229.
2. P. Rife, *Lise Meitner and the Dawn of the Nuclear Age*, p. 135.
3. J. Heilbron, *Dilemmas*, p. 180.
4. I make an "odd couple" reference with sincere apologies to Massimiliano Badino and his excellent article, "The Odd Couple: Boltzmann, Planck and the Application of Statistics to Physics (1900–1913)." *Annalen der Physik*, 18 (2009): 81–101. We agree that Planck is Felix, but I see Einstein as the better Oscar, especially in terms of an actual friendship with Planck/Felix.

5. M. Planck, Leden address of 1929, as in J. Heilbron, *Dilemmas*, p. 140.
6. A. Einstein, preface to M. Planck's, *Where Is Science Going?* pp. 9–13.
7. W. Isaacson, *Einstein*, pp. 21–23.
8. Ibid, pp. 8–9.
9. Ibid, pp. 16–17.
10. As in A. Folsing, *Albert Einstein: A Biography*, p. 243.
11. W. Isaacson, *Einstein*, p. 132.
12. A. Einstein to J. Laub, May 19, 1909. *The Collected Papers of Albert Einstein.*
13. D. Hoffmann, in *Annalen der Physik* 17 (2008), p. 285.
14. J. Heilbron, *Dilemmas*, p. 62.
15. M. Planck to H. Lorentz, October 21, 1898, as in A. J. Kox, *The Scientific Correspondence of Hendrik* Lorentz, p. 74.
16. M. Planck to H. Lorentz, January 17, 1899, as in A. J. Kox, *The Scientific Correspondence of Hendrik Lorentz*, p. 83.
17. A. Einstein to M. Solovine, April 27, 1906, as in W. Isaacson, *Einstein,* p. 141.
18. P. Frank, *Einstein: His Life and Times*, p. 190.
19. M. Planck, *Scientific Autobiography and Other Papers*, pp. 46–47.
20. A. Einstein. "On the Electrodynamics of Moving Bodies." *Annalen der Physik*, 17 (1905): 891.
21. To look at the second postulate more comprehensively, by declaring the constancy of light speed, Einstein was in some sense extending Galileo's statement of relativity to encompass not just *mechanics* but *electromagnetism* as well.
22. J. Heilbron, *Dilemmas*, p. 28.
23. Ibid.
24. M. Planck to A. Einstein, July 7, 1907, as in ibid.
25. D. Hoffmann, *Einstein's Berlin*, p. 137.
26. A. Roguin. "Christian Johann Doppler: The Man behind the Effect." *British Journal of Radiology*, 75 (2002): 615–619.
27. As in T. Levenson, *Einstein in Berlin*, pp. 421–422.
28. As in W. Isaacson, *Einstein*, p. 409.
29. F. Stern, *Einstein's German World*, p. 56.
30. As in W. Isaacson, *Einstein*, p. 408.
31. G. Pihl, *Germany: The Last Phase*, pp. 228–229.

CHAPTER 7

1. A. C. Grayling, *Among the Dead Cities*, p. 62.
2. As I edited this chapter, in January of 2014, a buried, long-dormant bomb exploded in a Euskirchen construction site, killing a worker. This type of "sleeper" bomb event is hardly uncommon.
3. W. G. Sebald, *On the Natural History of Destruction*, pp. 18–19.
4. Royal Air Force digital archives.

5. A. v. Pufendorf, *Die Plancks*, p. 442.
6. J. Heilbron, *Dilemmas*, p. 193.
7. M. Planck, *Scientific Autobiography and Other Papers*, pp. 78–79.
8. A. v. Pufendorf, *Die Plancks*, p. 17.
9. *Deutsches Geschlechterbuch, Band 178*, "Family peiper," p. 183.
10. M. Planck, *Scientific Autobiography*, p. 21; and T. Kuhn, *Black-Body Theory and the Quantum Discontinuity*, p. 17.
11. M. Planck, *Scientific Autobiography*, p. 21.
12. A. Hermann, *Planck*, pp. 18–19.
13. D. Hoffmann, *Quantum Theory Centenary*, p. 37.
14. Iris Runge quoted as in A. Hermann, *Planck*, pp. 11–13.
15. M. Planck, C. Runge, et al., *Brieftagebuch*, p. 105, entry of April 19, 1888.
16. M. Planck, *Scientific Autobiography and Other Papers*, p. 15.
17. M. Planck, C. Runge, et al., *Brieftagebuch*, p. 101, entry of June 6, 1886.
18. Ibid., p. 103, entry of October 9, 1886.
19. Ibid.
20. *Deutsches Geschlechterbuch, Band 178*, "Family peiper," p. 183.
21. M. Planck, C. Runge, et al., *Brieftagebuch*, p. 107, entry of July 15, 1888.
22. D. Hoffmann, *Quantum Theory Centenary*, p. 37; R. L. Sime. *Lise Meitner*, p. 13; D. Hoffmann, *Max Planck: Die Entstehung*, pp. 39–40.
23. J. Heilbron, *Dilemmas*, p. 15.
24. Ibid., p. 12.
25. M. Planck, *Scientific Autobiography*, pp. 33–34.
26. J. T. Blackmore, "Is Planck's 'Principle' True?" *The British Journal for the Philosophy of Science*, pp. 347–349.
27. M. Planck, C. Runge, et al., *Brieftagebuch*, p. 156, entry of October 14, 1905.
28. A. Hermann, *Planck*, p. 20.
29. M. Planck, C. Runge, et al., *Brieftagebuch*, p. 117, entry of August 7, 1890.
30. Ibid., p. 107, entry of July 15, 1888.
31. J. Heilbron, *Dilemmas*, p. 40.
32. M. Planck to A. Einstein, July 6, 1907. *The Collected Papers of Albert Einstein.*
33. M. Planck, C. Runge, et al., *Brieftagebuch*, pp. 173–174, entry of March 28, 1909.
34. Ibid., pp. 175–176, entry of September 1, 1909.
35. As in A. v. Pufendorf, *Die Plancks*, p. 39.
36. As in J. Heilbron, *Dilemmas*, p. 46.
37. J. Mehra and H. Rechenberg, *Schrödinger in Vienna and Zurich*, p. 97.
38. As in A. v. Pufendorf, *Die Plancks*, p. 39.
39. A. v. Pufendorf, *Die Plancks*, p. 41.
40. M. Planck, C. Runge, et al., *Brieftagebuch*, pp. 156–157, entry of October 14, 1905.
41. A. v. Pufendorf, *Die Plancks*, p. 43.
42. Ibid.

CHAPTER 8

1. A. v. Pufendorf, *Die Plancks*, p. 444.
2. Ibid.
3. "Suicides: Nazis go down to defeat in a wave of selbstmord." *Life* magazine, May 14, 1945.
4. C. Goeschel, *Suicide in Nazi Germany*.
5. Dr. G. Rabel as interviewed by T. Kuhn, 1963. American Institute of Physics Archives.
6. J. Heilbron, *Dilemmas*, p. 193.
7. A. v. Pufendorf, *Die Plancks*, p. 57.
8. Ibid., p. 60.
9. M. Planck, C. Runge, et al., *Brieftagebuch*, pp. 171–172, entry of October 18, 1908.
10. A. v. Pufendorf, *Die Plancks*, pp. 57–58; and J. Heilbron, *Dilemmas*, p. 64.
11. J. Heilbron, *Dilemmas*, p. 64.
12. J. Heilbron, *Dilemmas*, p. 72; A. v. Pufendorf, *Die Plancks*, p. 76.
13. M. Planck in *Sitzungsberichte*, January 23, 1913.
14. M. Planck, *Physikalische Abhandlungen und Vorträge*, as in J. Heilbron, *Dilemmas*, p. 65.
15. W. Isaacson, *Einstein: His Life and Universe*, p. 206.
16. M. Planck to E. and M. Lenz, September 17, 1914, and M. Planck to W. Wien, November 8, 1914, as in J. Heilbron, *Dilemmas*, p. 72.
17. P. Englund, *The Beauty and the Sorrow*, pp. 8–9.
18. Ibid., p. 86.
19. J. Heilbron, *Dilemmas*, p. 68.
20. A. Einstein to P. Ehrenfest, August 19, 1914, as in T. Levenson, *Einstein in Berlin*, p. 60.
21. A. Einstein, "The World as I See It," as in T. Levenson, *Einstein in Berlin*, p. 60.
22. L. Meitner to O. Hahn, November 16, 1916, as in P. Rife, *Lise Meitner and the Dawn of the Nuclear Age*, p. 64.
23. T. Levenson, *Einstein in Berlin*, pp. 60–61.
24. M. Planck, C. Runge, et al., *Brieftagebuch*, p. 189, entry of March 13, 1917.
25. P. Englund, *The Beauty and the Sorrow*, p. 66.
26. Ibid., p. 65.
27. Ibid.
28. Ibid., pp. 59–61.
29. Ibid., p. 92.
30. Ibid., p. 80.
31. M. Planck, C. Runge, et al., *Brieftagebuch*, pp. 188–189, entry of January 5, 1916.
32. A. v. Pufendorf, *Die Plancks*, p. 96; Planck to Lorentz, March 28, 1915, as in J. Heilbron, *Dilemmas*, p. 74.

33. M. Planck to M. and E. Lenz, July 24, 1916. Archiv der Max-Planck-Gesellschaft, Dept. Va, 11 Rep, No. 729, Max Planck collection.
34. A. v. Pufendorf, *Die Plancks,* p. 53.
35. Ibid., pp. 99–100; *Deutsches Geschlechterbuch Band 178*, "Family peiper," p. 183.
36. A. v. Pufendorf, *Die Plancks,* p. 94.
37. Ibid.
38. M. Planck, C. Runge, et al., *Brieftagebuch*, pp. 189–190, entry of March 13, 1917.
39. A. v. Pufendorf, *Die Plancks*, pp. 101–102.
40. P. Englund, *The Beauty and the Sorrow*, p. 254.
41. M. Planck to M. and E. Lenz, July 24, 1916. Archiv der Max-Planck-Gesellschaft, Dept. Va, 11 Rep, No. 729, Max Planck collection.
42. Planck to Wien, August 19, 1918, as in J. Heilbron, *Dilemmas*, p. 82.
43. R. L. Sime, *Lise Meitner*, pp. 39–40.
44. J. B. Scott, *Official statements of war aims and peace proposals, December 1916 to November 1918.* Pamphlet Series of the Carnegie Endowment for International Peace. 1921.
45. "Fate of Germany depends on the potato," *New York Times*, April 6, 1917.
46. P. Englund, *The Beauty and the Sorrow*, p. 329.
47. Dr. G. Rabel as interviewed by T. Kuhn, 1963. American Institute of Physics Archives.
48. W. Isaacson, *Einstein*, p. 239.
49. "Solving the Problem of Infant Mortality," *Harper's*, October 1917, pp. 723–729.
50. P. Englund, *The Beauty and the Sorrow*, footnote p. 471.
51. A. v. Pufendorf, *Die Plancks*, pp. 116–118.
52. Ibid., p. 122.
53. M. Planck to A. Einstein, December 29, 1917. *The Collected Papers of Albert Einstein.*
54. P. Englund, *The Beauty and the Sorrow*, p. 470.
55. A. Einstein to P. Ehrenfest, March 1, 1918, as in J. Heilbron, *Dilemmas*, p. 84.
56. M. Born to A. Einstein, October 23, 1920, as in J. Heilbron, *Dilemmas*, p. 85
57. S. Ansart et al. "Mortality burden of the 1918–1919 influenza pandemic in Europe." *Influenza and Other Respiratory Viruses*, 3 (2009): 99–106.
58. As in R. L. Sime, *Lise Meitner*, p. 73.
59. M. Planck, C. Runge, et al., *Brieftagebuch*, pp. 192–192, entry of December 28, 1918.
60. As in W. Isaacson, *Einstein*, p. 242.
61. R. L. Sime, *Lise Meitner*, p. 75.
62. J. M. Keynes, as quoted in H. M. Pachter, *Modern Germany*, p. 101.
63. E. Fischer to S. Arrhenius, May 12, 1919, as in J. Heilbron, *Dilemmas*, p. 101.
64. Nobel Prize Biography of M. Planck, Nobel Media AB.

65. D. Hoffmann, "'. . . you can't say to anyone to their face: your paper is rubbish.' Max Planck as Editor of the *Annalen der Physik*," *Annalen der Physik* 17 (2008), p. 280.
66. M. Planck to E. Wiedemann, June 8, 1919; M. Planck in *Berliner Tageblatt*, 25 December 1919, as in J. Heilbron, *Delimmas*, p. 88; and M. Planck, C. Runge, et al., *Brieftagebuch*, pp. 192–192, entry of December 28, 1918.
67. A. v. Pufendorf, *Die Plancks*, p. 156.
68. J. Heilbron, *Dilemmas*, p. 83.
69. A. Einstein to M. Born, December 9, 1919, as in J. Heilbron, *Dilemmas*, p. 83–84.
70. M. Planck to H. Lorentz, December 21, 1919, as in J. Heilbron, *Dilemmas*, p. 83.
71. M. Planck to E. Lenz, January 3, 1920, as in J. Heilbron, *Dilemmas*, p. 84.
72. M. Planck to Kippenberg, March 14, 1945, as in J. Heilbron, *Dilemmas*, pp. 195–196.
73. A. v. Pufendorf, *Die Plancks*, p. 156.

CHAPTER 9

1. A. v. Pufendorf, *Die Plancks,* p. 444.
2. Ibid., p. 445.
3. W. G. Sebald, *On the Natural History of Destruction*, pp. 4, 30, 33, 38.
4. Records of the Office of Strategic Services, RG 226. 42025, 44803, 45356. National Archives and Records Administration, Washington, DC.
5. As in G. A. Craig, *Germany: 1866–1945*, p. 762.
6. Hans-Georg von Studnitz, *Als Berlin brannte: Diarium der Jahre 1943–1945*, as in G. A. Craig, *Germany: 1866–1945*, p. 761.
7. P. Hoffmann, *The History of the German Resistance 1933–1945*, pp. 110, 270.
8. A. v. Pufendorf, *Die Plancks,* p. 172.
9. T. Levenson, *Einstein in Berlin,* p. 251.
10. J. Heilbron, *Dilemmas*, p. 87.
11. Ibid., p. 89.
12. Ibid., p. 90.
13. Ibid., p. 92.
14. K. Hentschel, *Physics and National Socialism,* Exhibit 3; J. Stark and P. Lenard, "The Hitler Spirit and Science," *Grossdeutsche Zeitung.* May 8, 1924.
15. J. Heilbron, *Dilemmas*, pp. 114–115.
16. Ibid.
17. M. Planck to A. Einstein, September 5, 1920, as in J. Heilbron, *Dilemmas*, p. 117.
18. K. Hentschel, *Physics and National Socialism*, Exhibit 1.
19. J. Heilbron, *Dilemmas*, p. 118.
20. W. Isaacson, *Einstein*, p. 287, endnote 19.
21. J. Heilbron, *Dilemmas*, p. 119.

22. W. Isaacson, *Einstein*, p. 288.
23. J. Heilbron, *Dilemmas*, p. 117, and W. Isaacson, *Einstein*, p. 288.
24. M. Eckert, *Arnold Sommerfeld*, p. 239.
25. J. Heilbron, *Dilemmas*, p. 119.
26. P. Lenard to W. Wien, June 6, 1922, August 25, 1922, and July 20, 1925. From the Deutches Museum Archives, Munich, collected letters of Wilhelm Wien.
27. K. Hentschel, *Physics and National Socialism*, Exhibit 3; J. Stark and P. Lenard, "The Hitler Spirit and Science," *Grossdeutsche Zeitung*, May 8, 1924.
28. T. Levenson, *Einstein in Berlin*, p. 250.
29. Ibid.
30. W. Isaacson, *Einstein*, p. 223.
31. A. Einstein to J. Laub. August 10, 1911. *The Collected Papers of Albert Einstein*.
32. T. Levenson, *Einstein in Berlin*, pp. 1–2.
33. M. Planck, W. Nernst, H. Rubens, E. Warburg, proposal of June 12, 1913. *The Collected Papers of Albert Einstein*.
34. T. Levenson, *Einstein in Berlin*, pp. 2–3.
35. J. Heilbron, *Dilemmas*, p. 32.
36. C. M. Will. "Henry Cavendish, Johann von Soldner, and the Deflection of Light." *American Journal of Physics*, 56 (1988): 413. For more information on the tests and vindication of general relativity, see Clifford M. Will's wonderfully readable book, *Was Einstein Right?*
37. T. Levenson, *Einstein in Berlin*, p. 45.
38. Ibid., pp. 34–35, via A. Pais, *Subtle Is the Lord*.
39. W. Isaacson, *Einstein*, p. 133.
40. Ibid., p. 196.
41. Ibid., p. 220.
42. It is important to note here that Einstein's final general relativity actually underlines the fundamental *difference* between a gravitational effect and the effect of just accelerating, in opposition to the spirit of his "equivalence principle." The difference is subtle and easily overlooked.

    To understand how the two work with one another, we must look to the idea of a local effect versus a global one. This is easy to do by considering Earth. We can treat Earth as flat as long as we stay in one neighborhood, and we can think of gravity as pulling everywhere downward, with parallel lines of gravitational effect—would you not agree that the pull of gravity in your living room is parallel to the pull of gravity in your kitchen? So too is the equivalence principle valid in this local domain. But as we take our view to a point far above our neighborhood, we can see the Earth is curved. Gravity pulls everywhere *inward* toward the center of the Earth. The once parallel lines of gravitational effect are now more like the spokes of an enormous bicycle wheel. No acceleration could create this global gravitational effect, and there is literally *not* a global equivalence principle—it can only fool people in small spaces like the interior of a rocket ship.

43. A. Pais, *Subtle Is the Lord*, and W. Isaacson, *Einstein*, p. 218.
44. M. Planck to A. Einstein, October 4, 1919. *The Collected Papers of Albert Einstein.*
45. T. Levenson, *Einstein in Berlin*, p. 228.
46. Ibid., p. 223.
47. D. Hoffmann, *Einstein's Berlin*, p. 137.
48. J. Heilbron, *Dilemmas*, p. 120.
49. Emma Planck to Erwin Planck, 1916, as in A. v. Pufendorf, *Die Plancks*, p. 110.
50. P. Frank, *Einstein: His Life and Times*, p. 192.
51. J. Heilbron, *Dilemmas*, p. 121. (Yes, Stark was writing of himself in a "royal" third person.)
52. K. Hentschel, *Physics and National Socialism*, Exhibit 2.
53. A. Einstein to M. Planck, July 6, 1922. *The Collected Papers of Albert Einstein.*
54. M. Planck to A. Einstein, July 8, 1922. *The Collected Papers of Albert Einstein.*
55. J. Heilbron, *Dilemmas*, p. 121.
56. Ibid.
57. R. L. Sime, *Lise Meitner*, pp. 109–110.
58. A. v. Pufendorf, *Die Plancks*, p. 192.
59. J. Heilbron, *Dilemmas*, p. 121.

CHAPTER 10

1. As in A. v. Pufendorf, *Die Plancks*, pp. 445–446.
2. J. Heilbron, *Dilemmas*, p. 3.
3. Ibid., p. 12.
4. T. Kuhn, *Black-Body Theory and the Quantum Discontinuity*, p. 23.
5. J. Heilbron, *Dilemmas*, pp. 14–15.
6. T. Kuhn, *Black-Body Theory and the Quantum Discontinuity*, p. 22.
7. Ibid., pp. 61–62.
8. M. Planck, *Scientific Autobiography*, p. 38.
9. H. Hertz. "Ueber die Einwirkung einer geradlinigen electrischen Schwingung auf eine benachbarte Strombahn." *Annalen der Physik* 270, no. 5 (1888): 155–170.
10. G. Kirchhoff. "Ueber das Verhältniss zwischen dem Emissionsvermögen und dem Absorptionsvermögen der Körper für Wärme and Licht." *Annalen der Physik und Chemie* 109, no. 2 (1860): 275–301.
11. In truth, even today, creating the experimental conditions necessary for a *perfect* black body are nearly impossible to achieve; on this basis, some have noted that Kirchhoff "over-reached" with his prescient claim.
12. M. Klein, "Max Planck and the Beginnings of Quantum Theory." *Archive for the History of Exact Sciences* 1, no. 5 (1962): 462.
13. As noted in T. Kuhn, *Black-Body Theory and the Quantum Discontinuity*, p. 34.

14. In the famous 9th edition of the *Encyclopaedia Britannica*, James Clerk Maxwell had written, under his "Atom" entry that, "the molecular motions" in a substance "are the source of the emitted light." Although Maxwell had written these words roughly 20 years before Planck entered the black-body problem, the full *Encyclopaedia* was only published in 1889, and in any case, I can find no evidence that this work was translated into German. Germany already had the well-regarded *Brockhaus Enzyklopädie*, so the market for a translated version from England would have been very small.
15. T. Kuhn, *Black-Body Theory and the Quantum Discontinuity*, pp. 33–37.
16. Ibid., pp. 76–77.
17. Ibid., pp. 70–71.
18. As in T. Kuhn, *Black-Body Theory and the Quantum Discontinuity*, p. 91.
19. M. Klein, "Max Planck and the Beginnings of the Quantum Theory." *Archive for the History of Exact Sciences 1*, no. 5 (1962): 464.
20. As in H. Kangro, *Planck's Original Papers in Quantum Physics*, p. 37.
21. W. Wien to M. Planck, November 16, 1900 and M. Planck to W. Wien, November 25, 1900. From the Deutches Museum Archives, Munich, collected letters of Wilhelm Wien.
22. M. Planck, *Scientific Autobiography*, p. 41.
23. M. Klein, "Max Planck and the Beginnings of the Quantum Theory." *Archive for the History of Exact Sciences 1*, no. 5 (1962): 468.
24. Ibid., p. 472.
25. As in H. Kangro, *Planck's Original Papers in Quantum Physics*, p. 40.
26. H. Kragh, "Max Planck: The Reluctant Revolutionary." *Physics World*, p. 33.
27. J. Heilbron, *Dilemmas*, p. 23.
28. As in H. Kangro, *Planck's Original Papers in Quantum Physics*, pp. 44–45.
29. J. Heilbron, *Dilemmas*, pp. 21–22.
30. M. Planck, *Scientific Autobiography*, p. 42
31. A. Einstein, in a 1947 open letter on behalf of the National Academy of Sciences. American Philosophical Society Archives.
32. T. Kuhn, *Black-Body Theory and the Quantum Discontinuity*, pp. 134–135.
33. M. Klein, "Max Planck and the Beginnings of the Quantum Theory." *Archive for the History of Exact Sciences 1*, no. 5 (1962): 476.
34. *The Collected Papers of Albert Einstein*, June 12, 1933. The joint letter was presented by Planck, Nernst, Rubens, and Warburg, but the sentiment is certainly Planck's.
35. As in M. Klein, "Max Planck and the Beginnings of the Quantum Theory." *Archive for the History of Exact Sciences 1*, no. 5 (1962): 477.
36. W. Isaacson, *Einstein*, pp. 312–313.
37. M. Planck, *Scientific Autobiography*, pp. 44–45.
38. I refer interested parties to the wonderful new book by A. Douglas Stone, *Einstein and the Quantum: The Quest of the Valiant Swabian*. Stone does not

diminish Planck so much as underline and elevate the many key milestones built by Einstein in the emergence of quantum theory.

39. T. Kuhn, *Black-Body Theory and the Quantum Discontinuity.* See, for instance, the work of Darrigol, Needell, and Gearheart cited in the bibliography.
40. C. A. Gearheart, "Planck, the Quantum, and Historians." *Physics in Perspective* 4 (2002): 170–215.
41. M. Klein, "Max Planck and the Beginnings of the Quantum Theory." *Archive for the History of Exact Sciences 1*, no. 5 (1962): 459–479.
42. J. Franck, "Max Planck 1858–1947." *Science* 107 (1948): 534–537. Franck claimed this was "probably 1903."
43. M. Planck, C. Runge, et al., *Brieftagebuch*, pp. 168–169.
44. The first reference to quantized time I can find dates to 1927, when Robert Levi published at article entitled *Théorie de l'action universelle et discontinue*; Einstein famously speculated about the loss of spacetime continuity in a 1936 articled entitled "*Physik und Realität*," but with great doubt that such a theory was really possible; and the first earnest attempt appears to be Hartland Snyder's "Quantized Space-Time," in the *Physical Review* of 1947, nearly 40 years after Planck's note to Runge.
45. T. Kuhn, *Black-Body Theory and the Quantum Discontinuity*, pp. 190–192
46. Ibid., pp. 134–139
47. J. Heilbron, in *Quantum Mechanics at the Crossroads*, p. 27.
48. M. v. Laue to A. Einstein, June 2, 1906. *The Collected Papers of Albert Einstein.*
49. A. Einstein, "Die Plancksche Theorie der Srahlung und die Theorie der spezifischen Wärme." *Annalen der Physik* 327, no. 1 (1907): 180–190.
50. W. Nernst, March, 1910, as in T. Kuhn, *Black-Body Theory and the Quantum Discontinuity*, p. 214.
51. N. Straumann, "On the First Solvay Conference of 1911." *European Physical Journal H* 36 (2011): 379–399.
52. A. Einstein to Besso. December 26, 1911. *The Collected Papers of Albert Einstein.*
53. J. Heilbron, *Dilemmas*, p. 31.
54. Ibid., p. 21.
55. Ibid., p. 27.
56. M. Planck, *The Theory of Heat Radiation*, 2nd ed., p. viii.
57. D. Hoffmann, *Max Planck: Annalen Papers*, pp. 731–735.
58. The number 75 courtesy M. Cardona and W. Marx, "Max Planck: A Conservative Revolutionary." *Percorsi* 24 (2008): 41.
59. See for instance R. M. Wald, *Quantum Field Theory in Curved Spacetime and Black Hole Thermodynamics.*
60. See for instance S. Carroll, *From Eternity to Here: The Quest for the Ultimate Theory of Time.*
61. W. Eberling and D. Hoffmann, *Über Thermodynamische Gleichgewichte von Max Planck.* Berlin: Verlag Harri Deutsch, 2008; citation analysis by M. Cardona and W. Marx, "Max Planck: A Conservative Revolutionary," pp. 39–61.

62. M. Planck, Leiden Lecture 1908. As in J. Heilbron, *Dilemmas*, pp. 47–50.
63. H. Kragh, "Preludes to Dark Energy: Zero-Point Energy and Vacuum Speculations." *Archive for History of Exact Sciences* 66 (2012): 199–240.
64. M. Planck interview with J. Murphy, in *Where Is Science Going?* p. 218.
65. Ibid., pp. 215–216.
66. M. Caspar, *Kepler*, pp. 36–38.
67. J. Murphy, preface to *Where Is Science Going?* p. 38.

CHAPTER II

1. A. v. Pufendorf, *Die Plancks*, p. 447.
2. Ibid., p. 21.
3. Ibid., p. 22.
4. Ibid.
5. Ibid., pp. 28–29.
6. BBC digital archives, "On this Day," July 20, 1944.
7. BBC digital archives, "On this Day," July 21, 1944.
8. F. v. Schlabrendorff, *The Secret War Against Hitler*, p. 304.
9. H. M. Pachter, *Modern Germany*, p. 263.
10. F. Stern, *Einstein's German World*, p. 57.
11. C. Fitzgibbon, *To Kill Hitler: The Officers' Plot July 1944*, p. 122; J. Heilbron, *Dilemmas*, p. 195.
12. P. Hoffmann, *The History of the German Resistance 1933–1945*, p. 522.
13. J. Heilbron, *Dilemmas*, p. 194.
14. P. Hoffmann, *The History of the German Resistance 1933–1945*, pp. 178, 270; and K. von Klemperer, *German Resistance Against Hitler*, p. 213.
15. A. v. Pufendorf, *Die Plancks*, p. 453.
16. Ibid., p. 315.
17. J. Heilbron, *Dilemmas*, p. 194. However, one should naturally suspect the content of Erwin's alleged "confessions" under Gestapo interrogation.
18. H. Höhne, *The Order of the Death's Head: The Story of Hitler's SS*, p. 128; P. Hoffmann, *The History of the German Resistance 1933–1945*, p. 27.
19. A. v. Pufendorf, *Die Plancks*, p. 381.
20. J. Heilbron, *Dilemmas*, pp. 2–3.
21. Ibid., p. 104.
22. Ibid., p. 36.
23. Ibid., p. 37.
24. F. Stern, *Einstein's German World*, pp. 39–40. The Kaiser won the day, despite academic resistance from Planck and others. A special new law was passed to allow the government to discipline professors, separate from the university administration, and Arons was forced from his post. He went on to a life of pro-labor politics.
25. J. Heilbron, *Dilemmas*, p. 70.

26. Copy of the Appeal in the *North American Review* 210 (1919): 284–287.
27. L. Zuckerman, *The Rape of Belgium*, pp. 28–30.
28. M. Planck to H. Lorentz, March 28, 1915, as in J. Heilbron, *Dilemmas*, p. 75
29. M. Planck to H. Lorentz, March 20, 1916, as in J. Heilbron, *Dilemmas*, pp. 76–78.
30. M. Planck to H. Lorentz, September 22, 1925, as in J. Heilbron, *Dilemmas*, p. 104.
31. As in J. Heilbron, *Dilemmas*, p. 31.
32. Ibid., p. 81; M. Planck to A. Einstein, December 29, 1917. *The Collected Papers of Albert Einstein*.
33. M. Planck to M. v. Laue, August 31, 1943, as in J. Heilbron, *Dilemmas*, p. 105.
34. The "Planck Succession" took a great deal of time because the top candidate Arnold Sommerfeld conducted protracted negotiations on terms and then elected to stay in Munich, before the eventual replacement Erwin Schrödinger also prolonged negotiations with the University of Berlin. See M. Eckert, *Arnold Sommerfeld*, pp. 295–299.
35. J. Heilbron, *Dilemmas*, pp. 110–111.
36. Ibid., p. 107. Max v. Laue and Arnold Sommerfeld attended with Planck as well. M. Eckert, *Arnold Sommerfeld*, p. 301.
37. J. Murphy, introduction to M. Planck's *Where Is Science Going?* p. 15.
38. A. Einstein to G. S. Viereck, *Saturday Evening Post*, October 26, 1929.
39. M. Planck, *Sitzungsberichte*, 1926, as in J. Heilbron, *Dilemmas*, p. 144.
40. A. Einstein in epilogue of M. Planck, *Where Is Science Going?* p. 203.
41. Ibid., p. 218.
42. M. Pachter, *Modern Germany*, pp. 118–119, 127.
43. Ibid., p. 134.
44. J. Murphy, introduction to *Where Is Science Going?* p. 38.
45. J. Heilbron, *Dilemmas*, p. 141.
46. K. Hentschel, *Physics and National Socialism*, p. xlvii.
47. A. Einstein to M. Planck, October 10, 1930, as in J. Heilbron, *Dilemmas*, p. 199.
48. "Cabinet of Monocles," *Time*, June 13, 1932.
49. J. Murphy, introduction to M. Planck, *Where Is Science Going?* pp. 15–17.
50. Ibid., p. 16.
51. A. Einstein, preface to M. Planck, *Where Is Science Going?* p. 14.
52. W. Isaacson, *Einstein*, p. 397.
53. Ibid., p. 396.
54. Ibid., pp. 398–399.
55. Ibid., p. 401.
56. H. A. Turner, *Hitler's Thirty Days to Power*, pp. 148–149.
57. W. Isaacson, *Einstein*, p. 404
58. J. Heilbron, *Dilemmas*, p. 155.
59. K. Hentschel, *Physics and National Socialism*, Exhibit 6.

60. M. Planck to A. Einstein, March 31, 1933, as in T. Levenson, *Einstein in Berlin*, p. 420.
61. W. Isaacson, *Einstein*, p. 408.
62. D. Hoffman, *Einstein's Berlin*, p. 35.
63. D. Hoffmann and M. Walker, "The German Physical Society Under National Socialism," *Physics Today* 52 (2004): 58.
64. Marga Planck to P. Ehrenfest, April 26, 1933, as in J. Heilbron, *Dilemmas*, p. 151.
65. K. Macrakis, *Surviving the Swastika*, p. 53.
66. K. Hentschel, *Physics and National Socialism*, Exhibit 6.
67. J. Heilbron, *Dilemmas*, p. 158.
68. J. Stark. *Nature* 133 (April 21, 1933).
69. See v. Laue to G. Mie, November 22, 1934, as in K. Hentschel *Physics and National Socialism*, Exhibit 34; A. Einstein to P. Ehrenfest, April 25, 1912. *The Collected Papers of Albert Einstein*.
70. K. Hentschel, *Physics and National Socialism*, Exhibit 18.
71. W. Isaacson, *Einstein*, p. 410
72. L. Meitner to O. Hahn, March 21, 1933, as in K. Hentschel, *Physics and National Socialism*, Exhibit 5.
73. L. Meitner to O. Hahn, 1948, as in J. Heilbron, *Dilemmas*, p. 208.
74. Harald Bohr to R. G. D. Richardson, 30 May 1933, as in J. Heilbron, *Dilemmas*, p. 152.
75. G. Rabel as interviewed by T. Kuhn, 1963. American Institute of Physics Archives.
76. M. Planck to A. Einstein, October 26, 1918. *The Collected Papers of Albert Einstein*.
77. K. Macrakis, *Surviving the Swastika*, pp. 57–58.
78. Minutes of May 11, 1933 meeting of KWG, as in J. Heilbron, *Dilemmas*, p. 159.
79. J. Heilbron, *Dilemmas*, p. 150.
80. T. Powers, *Heisenberg's War*, pp. 38–39.

CHAPTER 12

1. A. v. Pufendorf, *Die Plancks*, pp. 448–450.
2. Ibid.
3. Ibid., p. 451.
4. Ibid., pp. 452–454.
5. K. Macrakis, *Surviving the Swastika*, p. 59.
6. Ibid., p. 62
7. Ibid., p. 65.
8. Ibid.
9. K. Hentschel, *Physics and National Socialism*, Exhibit 30.
10. J. Heilbron, *Dilemmas*, p. 206

11. F. Stern, *Einstein's German World*, p. 55
12. Ibid., p. 56.
13. K. Hentschel, *Physics and National Socialism*, Exhibit 38.
14. J. Franck, "Max Planck 1858–1947," *Science*, 107 (1948): 534–537.
15. P. Rife, *Lise Meitner and the Dawn of the Nuclear Age*, p. 135
16. "The Last Stand," *New York Times*, January 12, 1936, as in J. Heilbron, *Dilemmas*, pp. 152–153.
17. K. Hentschel, *Physics and National Socialism*, Exhibit 46.
18. J. Heilbron, *Dilemmas*, pp. 169–170
19. Ibid., p. 168
20. K. Hentschel, *Physics and National Socialism*, Exhibit 55; J. Heilbron, *Dilemmas*, pp. 191–192.
21. M. v. Laue, "My development as a physicist," in P. P. Ewald, ed. *Fifty Years of X-Ray Diffraction*, p. 297.
22. As in R. L. Sime, *Lise Meitner*, p. 336.
23. As in J. Heilbron, *Dilemmas*, p. 182.
24. Ibid., pp. 172–173.
25. D. Hoffmann, H. Rössler, and G. Reuther, " 'Lachkabinett' und 'grosses Fest' der Physiker. Walter Grotrians 'physiaklischer Einakter' zu Max Plancks 80. Geburtstag," *Ber. Wissenschaftsfesch*, 33 (2010): 30–53.
26. K. Hentschel, *Physics and National Socialism*, Exhibit 54.
27. J. Heilbron, *Dilemmas*, p. 174.
28. F. Stern, *Einstein's German World*, pp. 56–57.
29. Nobel Prize Biography of Carl von Ossietzky, Nobel Media AB.
30. A. v. Pufendorf, *Die Plancks*, p. 33.
31. G. Pihl, *Germany: The Last Phase*, pp. 21–23.
32. K. Hentschel, *Physics and National Socialism*, Exhibit 64.
33. A. v. Pufendorf, *Die Plancks*, p. 457.
34. Ibid., p. 452.
35. F. v. Schlabrendorff, *Offiziere gegen Hitler*, p. 139, as in A. v. Pufendorf, *Die Plancks*, pp. 455–456.
36. A. v. Pufendorf, *Die Plancks*, p. 458.
37. Ibid., p. 459.
38. Ibid., p. 452.
39. ibid., p. 461.

CHAPTER 13

1. M. Planck to N. Planck, October 17, 1944, as in A. v. Pufendorf, *Die Plancks*, pp. 456–457.
2. A. v. Pufendorf, *Die Plancks*, p. 462.
3. J. Heilbron, *Dilemmas*, p. 130.
4. T. Kuhn, *Black-Body Theory and the Quantum Discontinuity*, pp. 28, 76.

5. M. Planck, *The Theory of Heat Radiation*, 2nd ed., 1913, p. viii.
6. T. Kuhn, *Black-Body Theory and the Quantum Discontinuity*, p. 229, Figure 3a.
7. M. Planck to H. Lorentz, March 31, 1918, as in J. Heilbron, *Dilemmas*, p. 129.
8. T. Kuhn, *Black-Body Theory and the Quantum Discontinuity*, p. 182.
9. G. N. Lewis, *Nature*, December 18, 1926, pp. 118, 2981.
10. W. Isaacson, *Einstein*, p. 327.
11. S. Weinberg, *Lectures on Quantum Mechanics*, p. 14.
12. J. Heilbron, *Dilemmas*, p. 130.
13. Ibid., p. 127.
14. W. Isaacson, *Einstein*, pp. 316–317.
15. Ibid., p. 331.
16. J. Heilbron, *Dilemmas*, p. 134.
17. M. Planck, C. Runge, et al., *Brieftagebuch*, pp. 168–169.
18. To be precise, the universe says h divided by 4π.
19. M. Planck, *Scientific Autobiography and Other Papers*, p. 143.
20. W. Heisenberg, *Physics and Beyond*, p. 77.
21. W. Isaacson, *Einstein*, p. 333.
22. S. Weinberg, *Lectures on Quantum Mechanics*, p. 95.
23. As in J. Heilbron, *Dilemmas*, p. 135
24. A. D. Stone, *Einstein and the Quantum*.
25. J. S. Bell, *Speakable and Unspeakable in Quantum Mechanics*, 2nd ed.
26. W. Isaacson, *Einstein*, p. 335.
27. M. Planck, *Where Is Science Going?* p. 107. It would be a great mistake to think a single statement of Planck's summarizes his many iterations of thinking, and it would be an even larger mistake to take his version of causality as universally embraced. The discussion of causality has deservedly consumed many books in its own right, but here, we will stick close to Dr. Planck.
28. W. Isaacson, *Einstein*, p. 333.
29. Ibid.
30. M. Planck, *Where Is Science Going?* p. 110.
31. I am liberally summarizing about 70 pages of essay material dating to 1933. M. Planck, *Where Is Science Going?* pp. 107, 132, 141, 163.
32. J. Heilbron, *Dilemmas*, pp. 146–147.
33. M. Planck, "The Meaning and Limits of Exact Science." *Scientific Autobiography and Other Papers*, pp. 100–101.
34. M. Planck, "The Concept of Causality in Physics." *Scientific Autobiography and Other Papers*, p. 133.
35. M. Planck, as in J. Heilbron, *Dilemmas*, pp. 216–217.
36. J. Heilbron, *Dilemmas*, pp. 142–143.
37. M. Planck, *Scientific Autobiography and Other Papers*, pp. 132, 127, and 149.
38. A. v. Pufendorf, *Die Plancks*, p. 462.
39. Ibid.

40. M. Planck to H. and A. Hartmann, January 3, 1945. Archiv der Max-Planck-Gesellschaft, Dept. Va, 11 Rep, No. 1400, Max Planck collection.

CHAPTER 14

1. Plötzensee Memorial Center Hüttingpfad, D-13627, Berlin.
2. A. v. Pufendorf, *Die Plancks*, p. 465.
3. Marga Planck to M. von Laue, March 18, 1945, as in J. Heilbron, *Dilemmas*, p. 195.
4. As in A. v. Pufendorf, *Die Plancks*, p. 465.
5. M. Planck to Fritz and Grete Lenz, February 2, 1945, as in A. v. Pufendorf, *Die Plancks*, p. 466.
6. M. Planck to A. Sommerfeld, February 4, 1945, as in J. Heilbron, *Dilemmas*, p. 195.
7. M. Planck to Kippenberg, March 14, 1945, as in J. Heilbron, *Dilemmas*, p. 196.
8. A. Einstein to M. Planck, April 6, 1933, as in J. Heilbron, *Dilemmas*, p. 159.
9. K. Hentschel, *Physics and National Socialism*, p. liii.
10. Ibid., p. xlvii.
11. Ibid., Exhibit 9.
12. Ibid., Exhibit 10.
13. P. Rife, *Lise Meitner and the Dawn of the Nuclear Age*, pp. 111–112.
14. F. Haber to Rust, April 30, 1933, as in J. Heilbron, *Dilemmas*, p. 161.
15. K. Macrakis, *Surviving the Swastika*, p. 57.
16. J. Heilbron, *Dilemmas*, p. 210
17. As in T. Levenson, *Einstein in Berlin*, p. 420.
18. As in J. Heilbron, *Dilemmas*, p. 211
19. A. Hitler to E. Hepp, February 5, 1914, as in T. Levenson, *Einstein in Berlin*, p. 80.
20. A. Hitler, *Mein Kampf*, p. 202.
21. F. L. Coolidge, F. L. Davis, and D. L. Segal, "Understanding Madmen: A *DSM-IV* Assessment of Adolf Hitler." *Individual Differences Research*, 5 (2007): 30–43.
22. T. Mann, *Diaries 1918–1939*, Entry of December 20, 1934.
23. K. Hentschel, *Physics and National Socialism*, Exhibit 21, "The Scientific Situation in Germany," *Science*, 77 (June 2, 1933): 528–529.
24. K. Macrakis, *Surviving the Swastika*, pp. 57–58.
25. M. Planck to G. Rabel, May 24, 1933, as in J. Heilbron, *Dilemmas*, p. 214.
26. E. Larson, *In the Garden of Beasts*, pp. 76–79.
27. As in T. Levenson, *Einstein in Berlin*, p. 423.
28. J. James, T. Steinhauser, D. Hoffmann, and B. Friedrich, *100 Years at the Intersection of Chemistry and Physics*, p. 14.

29. K. Macrackis, *Surviving the Swastika*, p. 59. M. Planck, report of the MPG President.
30. As in T. Levenson, *Einstein in Berlin*, pp. 421–422. A. Einstein to M. Born, 1933.
31. K. Hentschel, *Physics and National Socialism*, Exhibit 22, W. Heisenberg to M. Born, June 2, 1933.
32. M. Eckert, *Arnold Sommerfeld*, p. 353.
33. J. Heilbron, "Max Planck's Compromises," in *Quantum Mechanics at the Crossroads*, p. 33.
34. As in J. Heilbron, *Dilemmas*, p. 206.
35. M. Planck, *Scientific Autobiography and Other Papers*, pp. 24–25.
36. Ibid., p. 23.
37. J. Heilbron, *Dilemmas*, pp. 23–25.
38. M. Planck, *Scientific Autobiography and Other Papers*, p. 33.
39. As in J. Heilbron, *Dilemmas*, pp. 205–207.
40. A. v. Pufendorf, *Die Plancks*, p. 456.
41. Ibid., pp. 463–464.
42. F. v. Schlabrendorff, *The Secret War Against Hitler*, p. 313.
43. A. v. Pufendorf, *Die Plancks*, p. 464.

CHAPTER 15

1. J. Granberg. "Berlin, Nerves Racked by Air Raids, Fears Russian Army Most." *Oakland Tribune*, February 23, 1945; F. v. Schlabrendorff, *The Secret War Against Hitler*, p. 325
2. As in A. v. Pufendorf, *Die Plancks*, pp. 465–466.
3. Ibid., p. 468.
4. G. A. Craig, *Germany: 1866–1945*, p. 759.
5. This broadcast is available via, for instance, "Old Time Radio": http://www.otr.com.
6. Dachau Concentration Camp Memorial Site, Alte Römerstrasse 75, D-85221 Dachau, DE.
7. D. Hoffmann, *Max Planck: Die Entstehung*, p. 105.
8. J. Heilbron, *Dilemmas*, p. 196.
9. B. B. Miltonberger and J. A. Huston, *134th Infantry Regiment Combat History of World War II*, Chapter 11.
10. M. Planck, C. Runge, et al., *Brieftagebuch*, pp. 173–174.
11. D. Cahan, "Helmholtz and the Ideals of Science and Culture in Gilded Age America." *Revista da SBHC*, 4 (2006): 6–16.
12. G. A. Kimble and M. Wertheime, *Portraits of Pioneers in Psychology*, vol. IV, p. 26.
13. M. Planck to M. v. Laue, April 29, 1909. Deutsches Museum Archives, Munich, Ref. 1951-8.
14. As in J. Heilbron, *Dilemmas*, p. 51.

15. R. Rosenzweig and E. Blackmar, *The Park and the People: A History of Central Park*. Ithaca, NY: Cornell University Press, 1992.
16. Coursework in German and scientific German were required for American physics students well into the 1950s.
17. M. Planck, *Eight Lectures on Theoretical Physics*, p. 2.
18. Ibid., p. 6.
19. J. Heilbron, *Dilemmas*, p. 54.
20. M. Planck, *Eight Lectures on Theoretical Physics*, p. 19.
21. As in J. Heilbron, *Dilemmas*, p. 40.
22. M. Planck, *Eight Lectures on Theoretical Physics*, pp. 95–96.
23. Ibid., p. 120.
24. Ibid., pp. 129–130.
25. M. Planck, C. Runge, et al., *Brieftagebuch*, pp. 175–176.
26. M. Planck's Rector's address of November 1, 1913, as in J. Heilbron, *Dilemmas*, p. 64.
27. F. C. Pogue, *The Supreme Command*. Washington, DC: Office of the Chief of Military History, 1954, pp. 479–491.
28. A. Beevor, "They Raped Every German Female from Eight to 80." *The Guardian*, April 30, 2002.

CHAPTER 16

1. As in C. G. Lasby, *Project Paperclip*, p. 105.
2. K. Lowe, *Savage Continent*, pp. 24, 27.
3. J. Heilbron, *Dilemmas*, p. 196.
4. S. A. Goudsmit, *Alsos*, p. 26
5. Ibid., pp. 47–49.
6. Ibid., p. 8.
7. Ibid., p. 71.
8. Ibid., pp. 105–106.
9. Ibid., p. 84.
10. H. B. G. Casimir, *Haphazard Reality: Half a Century of Science*, p. 281; also, as relayed in A. Hermann's *Max Planck*.
11. A. v. Pufendorf, *Die Plancks*, p. 469.
12. Ibid., p. 463.
13. R.V. Jones, foreword to *Alsos*, p. xiv. Jones claims to be the originator of the Farm Hall idea.
14. *Operation Epsilon*, p. 33.
15. S. Goudsmit, *Alsos*, p. 134.
16. *Operation Epsilon*, p. 73.
17. S. Goudsmit, *Alsos*, p. 135.
18. T. Powers, *Heisenberg's War*, pp. 95–96.
19. As in J. Heilbron, *Dilemmas*, p. 203.

CODA

1. *Max Planck und die Max-Planck-Gesellschaft.* Berlin: Archiv der Max-Planck-Gesellschaft, 2009, p. 265.
2. A. v. Pufendorf, *Die Plancks*, p. 468.
3. As in D. Hoffmann, *Max Planck*, p. 107.
4. J. Heilbron, *Dilemmas*, p. 1.
5. Marga Planck to F. Clarke, October 13, 1946. Archives of the American Philosophical Society.
6. M. Planck to Neuberg, September 17, 1946, as in J. Heilbron, *Dilemmas*, p. 197.
7. As in A. v. Pufendorf, *Die Plancks*, p. 469.
8. Obituary of Professor Max Planck, *The Times*, October 6, 1947. As in G. H. Markl and E. Henning (eds.), *Max-Planck-Gesellschaft Berichte und Mitteilungen*, Heft 3/97, p. 164.
9. As in J. Heilbron, *Dilemmas*, p. 197.
10. The *New York Times*, July 16, 1946.
11. As in D. Hoffmann, *Max Planck: Die Entstehung*, p. 110.
12. L. Meitner to M. Bohr, August 17, 1946, as in R. L. Sime, *Lise Meitner*, p. 334.
13. O. Hahn to L. Meitner, December 28, 1946, as in R. L. Sime, *Lise Meitner*, p. 344.
14. A. Westgren, Award Ceremony Speech of December 10, 1945, for the Nobel Prize in Chemistry. Nobel Media AB.
15. K. Hentschel, *Physics and National Socialism*, Exhibit 117, M. v. Laue. "The Wartime Activities of German Scientists," in *The Bulletin of Atomic Scientists*, 1948.
16. K. Hentschel, *Physics and National Socialism*, Exhibit 120, L. Meitner to O. Hahn, June 6, 1948.
17. As in R. L. Sime, *Lise Meitner*, pp. 338–339.
18. G. H. Markl and E. Henning (eds.), *Max-Planck-Gesellschaft Berichte und Mitteilungen*, Heft 3/97, p. 14.
19. *Max Planck Revolutionär wider Willen.* Heidelberg: Spektrum der Wissenschaft Verlagsgesellschaft mbH, 2008, p. 80.
20. J. Heilbron, "Max Planck on the way to and from the Absolute," pp. 28–29; D. Hoffmann, *Max Planck*, pp. 116–117.
21. Marga Planck to L. Meitner, June 15, 1947, as in "Max Planck: Life, Work, Personality. On the 50th Anniversary of His Death," Archiv der Max-Planck-Gesellschaft Exhibit at the Magnus-Haus, Berlin, 1997. See web archive: http://www.max-planck.mpg.de.
22. G. H. Markl and E. Henning, *Max Planck (1858–1947): Zum Gedenken an seinen 50. Todestag am 4. Oktober 1997*, p. 14.
23. G. Rabel as interviewed by T. Kuhn, 1963. American Institute of Physics Archives.
24. A. v. Pufendorf, *Die Plancks*, p. 470.
25. M. v. Laue, from the introduction to M. Planck's *Scientific Autobiography*, p. 11.

26. As in A. Hermann, *Planck*, p. 126.
27. M. Planck, *Where Is Science Going?* p. 83.

APPENDIX

1. As in H. Kangro's *Planck's Original Papers in Quantum Physics.*
2. See for instance: D. A. Stone, *Einstein and the Quantum*; D. Kleppner, "Rereading Einstein on Radiation." *Physics Today* (February 2005): 30–33. Einstein's 1917 approach introduced his "A and B coefficients," for comparing the rates of spontaneous (unprovoked) versus stimulated (caused by the absorption of incoming photons) emission of photons. Although the A and B coefficients arguably deserve their own appendix, and the approach led eventually to the laser, the topic strays too far from Dr. Planck.
3. It is important for students of physics to note that, although our text here opines about the universal nature of thermal radiation, it is not so simple in laboratory experiments. In fact, most objects function like "gray bodies" given various surface imperfections that keep them from becoming ideal black bodies. The fact that Planck's professor Kirchhoff faced reams of laboratory data that showed *different* spectra yet proposed a universal underlying mechanism, is one of the more incredible and arguably unsung parts of the story. Kirchhoff coined the term "gray bodies" even as he wrote of black bodies as well. To understand the extent to which lab measurements do *not* show ideal black-body behavior, see, for instance: P.-M. Roitaille, "Kirchoff's Law of Thermal Emission: 150 Years," *Progress of Physics*, 4 (2009): 3–13.
4. M. Planck, *Vorlesungen über die Theorie der Wärmestrahlung*, 1913.
5. T. Kuhn, *Black-Body Theory and the Quantum Discontinuity*, p. 253.
6. H. Kragh, *Archive for the History of Exact Sciences*, 66 (2012): 199–240.

Bibliography

Ansart, S., et al. "Mortality Burden of the 1918–1919 Influenza Pandemic in Europe." *Influenza and Other Respiratory Viruses* 3 (2009): 99–106.

Badino, Massimiliano. "The Odd Couple: Boltzmann, Planck and the Application of Statistics to Physics (1900–1913)." *Annalen der Physik* 18 (2009): 81–101.

Beck, Lorenz Friedrich, and Marion Kazemi. *Max Planck und die Max-Planck-Gesellschaft: Veröffentlichungen aus den Archiv der Max-Planck-Gesellschaft, Band 20.* Berlin: Archiv der MPG, 2009.

Bell, John S. *Speakable and Unspeakable in Quantum Mechanics*, 2nd ed. Cambridge: Cambridge University Press, 2004.

Blackmore, John T. "Is Planck's 'Principle' True?" *The British Journal for the Philosophy of Science* 29 (1978): 347–349.

Born, Max. "Max Karl Ernst Ludwig Planck. 1858–1947." *Biographical Memoirs of the Fellows of the Royal Society* 6, no. 17 (1948): 161–188.

Brush, Stephen G. "Cautious Revolutionaries: Maxwell, Planck, Hubble." *American Journal of Physics* 70 (2002): 119–127.

Cahan, David. "Helmholtz and the Ideals of Science and Culture in Gilded Age America." *Revista da SBHC* 4 (2006): 6–16.

Cardona, Manuel, and Werner Marx. "Max Planck—A Conservative Revolutionary." *Percorsi* 24 (2008): 39–54.

Carroll, Sean. *From Eternity to Here: The Quest for the Ultimate Theory of Time.* New York: Dutton, 2010.

Casimir, Hendrik B. G. *Haphazard Reality: Half a Century of Science.* New York: Harper Collins, 1983.

Cassidy, David. *Uncertainty: The Life and Science of Werner Heisenberg.* New York: WH Freeman, 1991.

Clausius, Rudolf. *The Mechanical Theory of Heat*, 2nd ed. Translated by Walter R. Browne. London: Macmillan, 1879.

Cooke, James Francis. *Great Pianists on Piano Playing: Study Talks with Foremost Virtuosos.* New York: Dover, 1999.

Coolidge, Frederick L., Felicia L. Davis, and Daniel L. Segal. "Understanding Madmen: A *DSM-IV* Assessment of Adolf Hitler." *Individual Differences Research* 5, no. 007 (2007): 30–43.

Craig, Gordon A. *Germany 1866–1945.* Oxford: Oxford University Press, 1980.

Cruikshank, Dale P. "Gerard Peter Kuiper: 1905–1973, a Biographical Memoir." In *Biographical Memoirs*. Washington DC: National Academy of Sciences, 1993, 259–295.

Darrigol, O. *From c-Numbers to q-Numbers: The Classical Analogy in the History of Quantum Theory*. Berkeley, CA: University of California Press, 1992.

Ebeling, Werner, and Dieter Hoffmann (forward). *Über Thermodynamische Gleichgewichte von Max Planck*. Frankfurt am Main: Harri Deutsch, 2008.

Eckert, Michael. *Arnold Sommerfeld: Science, Life, and Turbulent Times 1868–1951*. New York: Springer, 2013.

The Collected Papers of Albert Einstein. Edited by J. Stachel, R. Schulmann, D. Kormos-Buchwald, et al. Princeton, NJ: Princeton University Press, 1998 edition.

Englund, Peter. *The Beauty and the Sorrow: An Intimate History of the First World War*. New York: Vintage Books, 2012.

Flamm, Dieter. "Ludwig Boltzmann—A Pioneer of Modern Physics," presented at the 20th Anniversary Congress of the History of Science (1997), arXiv:physics/9710007v1.

Fölsing, Albrecht. *Albert Einstein: A Biography*. Translated by Ewald Osers. New York: Penguin Books, 1997.

Frank, Charles. *Operation Epsilon: The Farm Hall Transcripts*. Berkeley, CA: University of California Press, 1993.

Frank, Philip. *Einstein: His Life and Times*. New York: Alfred A. Knopf, 1947.

Gearheart, Clayton A. "Einstein Before 1905: The Early Papers on Statistical Mechanics." *American Journal of Physics* 58 (1990): 468–480.

Gearhart, Clayton A. "Planck, the Quantum, and the Historians." *Physics in Perspective* 4 (2002): 170–215.

Goeschel, Christian. *Suicide in Nazi Germany*. New York: Oxford University Press, 2009.

Goudsmit, Samuel A. *The History of Modern Physics, Volume I: Alsos*. Los Angeles: Tomash, 1983.

Grayling, A. C. *Among the Dead Cities: The History and Moral Legacy of the WWII Bombing of Civilians in Germany and Japan*. New York: Walker, 2006.

Groves, Leslie R. *Now it Can be Told: The Story of the Manhattan Project*. New York: Harper and Brothers, 1962.

Hartmann, Hans. *Max Planck als Mensch und Denker*. Taschenbuch: Ullstein, 1964.

Heald, Mark A. "Where Is the 'Wien Peak'?" *American Journal of Physics* 71 (2003): 1322.

Heilbron, John. *The Dilemmas of an Upright Man: Max Planck and the Fortunes of German Science*. Cambridge, MA: Harvard University Press, 1996.

Heilbron, John. "Max Planck's compromises on the way to and from the Absolute." In *Quantum Mechanics at the Crossroads*. New York: Springer, 2007, pp. 21–37.

Heisenberg, Werner. *Physics and Beyond: Encounters and Conversations.* London: George Allen and Unwin, 1971.

Helmholtz, Hermann. *Popular Lectures on Scientific Subjects.* Translated by E. Atkinson. London: Longman, Green, 1881.

Hentschel, Klaus. *Physics and National Socialism: An Anthology of Primary Sources.* Berlin: Birkhauser, 1996.

Hentschel, K., and R. Tobies, eds. *Brieftagebuch swischen Max Planck, Carl Runge, Bernhard Karsten und Adolf Leopold.* Berlin: ERS-Verlag, 2003.

Hermann, Armin. *Max Planck. Mit Selbszeugnissen und Bilddokumenten.* Hamburg: Rowohlt Taschenbuch Verlag, 1973.

Hertz, Heinrich. *Electric Waves: Being Researches on the Motion of Electric Action with Finite Velocity through Space.* Translated by D. E. Jones. New York: Macmillan, 1893.

Hoffmann, Dieter. "Between Autonomy and Accommodation: The German Physical Society during the Third Reich." *Physics in Perspective* 7, no. 3 (2005): 293–329.

Hoffmann, Dieter. "'. . . you can't say to anyone to their face: your paper is rubbish.' Max Planck as Editor of the Annalen der Physik." *Annalen der Physik* 17, no. 5 (2008): 273–301.

Hoffmann, Dieter. *Max Planck: Die Entstehung der modernen Physik.* München: Verlag, 2008.

Hoffmann, Dieter. *Einstein's Berlin: In the Footsteps of a Genius.* Baltimore, MD: Johns Hopkins University Press, 2013.

Hoffmann, Dieter, ed. *Max Planck: Annalen Papers.* Weinheim: Verlag GmbH, 2008.

Hoffmann, Dieter, ed. *Max Planck und die moderne Physik.* Heidelberg: Springer-Verlag, 2010.

Hoffmann, Dieter, and M. Walker. "The German Physical Society under National Socialism." *Physics Today* (December 2004): 52–58.

Hoffmann, Dieter, and Mark Walker, eds. *The German Physical Society in the Third Reich.* Cambridge: Cambridge University Press, 2011.

Hoffmann, Dieter et al. *Quantum Theory Centenary.* Berlin: Deutsche Physikalische Gesellschaft, 2000, pp. 19–30.

Hoffmann, Peter. *The History of the German Resistance 1933–1945.* Cambridge, MA: MIT Press, 1977.

Höhne, Heinz. *The Order of the Death's Head: The Story of Hitler's SS.* New York: Coward-McCann, 1970.

Isaacson, Walter. *Einstein: His Life and Universe.* New York: Simon and Schuster, 2007.

James, J., T. Steinhauser, D. Hoffmann, and B. Friedrich. *100 Years at the Intersection of Chemistry and Physics: The Fritz Haber Institute of the Max Planck Society 1911–2011.* Berlin: de Gruyter, 2011.

Jannsen, Michel. "Reconsidering a Scientific Revolution: The Case of Einstein *versus* Lorentz." *Physics in Perspective* 4 (2002): 421–446.

Jungnickel, Christa. *Intellectual Mastery of Nature. Theoretical Physics from Ohm to Einstein, Volume 2.* Chicago, IL: University of Chicago Press, 1986.

Kangro, Hans. "Planck, Max Karl Ernst Ludwig." In *Complete Dictionary of Scientific Biography.* New York: Charles Scribner's Sons, 2008.

Kant, Immanuel. *Critique of Pure Reason.* New York: Penguin Classics, 2007 (first published in German in 1781).

Kimble, Gregory A., and Michael Wertheimer. *Portraits of Pioneers in Psychology, Volume IV.* New York: Taylor and Francis, 2000.

Kjellén, Rudolf. *Der Staat als Lebensform.* 1917. Reprint. London: Forgotten Books, 2013.

Klein, Martin. "Max Planck and the Beginnings of the Quantum Theory." *Archive for History of Exact Sciences* 1, no. 5 (1961): 459–479.

Klein, Martin. "Thermodynamics and Quanta in Planck's Work." *Physics Today* 19, no. 11 (1966): 23–28.

Klemperer, Klemens v. *German Resistance Against Hitler: The Search for Allies Abroad.* New York: Oxford University Press, 1994.

Kox, A. J., ed. *The Scientific Correspondence of H. A. Lorentz, Volume I.* New York: Springer, 2008.

Kragh, Helge. "Max Planck: the reluctant revolutionary." *Physics World* (December 2000): 31–35.

Kragh, Helge. "Preludes to dark energy: Zero-point energy and vacuum speculations." *Archive for the History of Exact Sciences* 66 (2012): 199–240.

Kuhn, Thomas. *Black-Body Theory and the Quantum Discontinuity 1894–1912.* New York: Oxford University Press, 1978.

Larson, Erik. *In the Garden of Beasts: Love, Terror, and an American Family in Hitler's Berlin.* New York: Crown Publishing, 2011.

Lasby, Clarence G. *Project Paperclip: German Science and the Cold War.* New York: Athenium, 1971.

Laue, Max v. "My Development as a Physicist." In *Fifty Years of X-Ray Diffraction,* edited by P. P. Ewald. Utrecht: International Union of Crystallography, 1962, pp. 278–307.

Lemmerich, Jost. *Science and Conscience: The Life of James Franck.* Translated by A. Hentschel. Stanford, CA: Stanford University Press, 2007.

Levenson, Thomas. *Einstein in Berlin.* New York: Bantam, 2003.

Lowe, Keith. *Savage Continent: Europe in the Aftermath of World War II.* New York: St. Martin's Press, 2012.

Lummer, O., and E. Pringsheim. "Die Verteilung der Energie im Spektrum des schwarzen Körpers." *Verh. Dt. Phys. Ges.* 1 (1899): 23–41.

Lummer, Otto. *Grundlagen, Ziele und Grenzen der Leuchttechnik (Auge und Lichterzeugung).* Munich, 1918.

Macrakis, Kristie. *Surviving the Swastika: Scientific Research in Nazi Germany*. Oxford: Oxford University Press, 1993.

Mann, Thomas. *Buddenbrooks*. New York: Alfred A. Knopf, 1924.

Mann, Thomas. *Diaries, 1918–1939*. Translated by Richard and Clara Winston. New York: H. N. Abrams, 1982.

Markl, G. H., and E. Henning. *Max Planck (1858–1947): Zum Gedenken an seinen 50. Todestag am 4. Oktober 1997*. München: Max Planck Gesellschaft, 1997.

Mehra, Jagdish, and Helmut Rechenberg. *Schrödinger in Vienna and Zurich*. New York: Springer Verlag, 2000.

Miltonberger, Butler B., and James A. Huston. *134th Infantry Regiment Combat History of World War II*. Baton Rouge, LA: Army and Navy Publishing Company, 1946.

Nagle, John F. "In Defense of Gibbs and the Traditional Definition of the Entropy of Distinguishable Particles." *Entropy* 12 (2010): 1936–1945.

Norton, John D. "'Nature is the Realisation of the Simplest Conceivable Mathematical Ideas': Einstein and the Canon of Mathematical Simplicity." *Studies in History and Philosophy of Modern Physics* 31 (2000): 135–170.

Pachter, Henry Maximilian. *Modern Germany: A Social, Cultural and Political History*. Boulder, CO: Westview Press, 1979.

Pais, Abraham. *Subtle Is the Lord: The Science and the Life of Albert Einstein*. New York: Oxford University Press, 2005.

Peierls, Rudolf. *Bird of Passage: Recollections of a Physicist*. Princeton, NJ: Princeton University Press, 1988.

Pihl, Gunnar. *Germany: The Last Phase*. New York: A. A. Knopf, 1944.

Planck, Max. *Eight Lectures on Theoretical Physics, Delivered at Columbia University in 1909*. New York: Columbia University Press, 1915.

Planck, Max. *Scientific Autobiography and Other Papers*. New York: Philosophical Library, 1949.

Planck, Max. *Treatise on Thermodynamics*. Translated from Planck's seventh edition, 1922, by Alexander Ogg. New York: Dover, 1969.

Planck, Max. "On an Improvement of Wien's Equation for the Spectrum." In *Classic Papers in Physics*, edited by Hans Kangro. London: Taylor and Francis, 1972, pp. 35–37.

Planck, Max. "On the Theory of the Energy Distribution Law of the Normal Spectrum." In *Classic Papers in Physics*, edited by Hans Kangro. London: Taylor and Francis, 1972.

Planck, Max. *Where Is Science Going?* Woodbrige: Oxbow, 1981.

Planck, Max. *The Theory of Heat Radiation*. Translated from Planck's 1913 manuscript, *Vorlesungen über die Theorie der Wärmestrahlung*, by Morton Masius. London: Forgotton Books, 2012.

Powers, Thomas. *Heisenberg's War: The Secret History of the German Bomb*. New York: Little, Brown, 1993.

Pufendorf, Astrid v. *Die Plancks: Eine Familie Zwischen Patriotismus Und Widerstand*. Berlin: Propyläen Verlag, 2006.

Rife, Patricia. *Lise Meitner and the Dawn of the Nuclear Age*. Boston: Birkhäuser, 1999.

Roguin, A. "Christian Johann Doppler: The Man Behind the Effect." *British Journal of Radiology* 75 (2002): 615–619.

Rosenzweig, R. and E. Blackmar. *The Park and the People: A History of Central Park*. Ithaca, NY: Cornell University Press, 1992.

Schlabrendorff, Fabian v. *Offiziere gegen Hitler*. Berlin: Siedler, 1984.

Sebald, W. G. *On the Natural History of Destruction*. New York: Modern Library, 2004.

Sime, Ruth Lewin. *Lise Meitner: A Life in Physics*. Berkeley, CA: University of California Press, 1996.

Soffer, B. H., and D. K. Lynch. "Some Paradoxes, Errors, and Resolutions Concerning the Spectral Optimization of Human Vision." *American Journal of Physics* 67, no. 11 (1999): 946–953.

Stern, Fritz. *Einstein's German World*. Princeton, NJ: Princeton University Press, 2001.

Stone, A. Douglas. *Einstein and the Quantum: The Quest of the Valiant Swabian*. Princeton, NJ: Princeton University Press, 2013.

Straumann, Norbert. "On the First Solvay Conference of 1911." *European Physical Journal H* 36 (2011): 379–399.

Turner, Henry A. *Hitler's Thirty Days to Power*. Reading, MA: Addison-Wesley, 1996.

Ullmann, Dirk. *Quelleninventar Max Planck*. Veröffentlichungen aus dem Archiv der Max-Planck-Gesellschaft, Band 8. Berlin: Archiv der MPG, 1996.

Ullmann, Dirk. "Max Planck als Wissenschaftsorganisator im Spiegel der archivalischen Überlieferung." *Physikalische Blätter* 53 (2013): 1017–2018.

Wald, Robert M. *Quantum Field Theory in Curved Spacetime and Black Hole Thermodynamics*. Chicago, IL: University of Chicago Press, 1994.

Walker, Mark. *Nazi Science: Myth, Truth, and the German Atomic Bomb*. New York: Perseus, 1995.

Weinberg, Steven. *Lectures on Quantum Mechanics*. Cambridge: Cambridge University Press, 2012.

Will, Clifford M. "Henry Cavendish, Johann von Soldner, and the Deflection of Light." *American Journal of Physics* 56 (1988): 413.

Wilson, Robert. "Discovery of the Cosmic Background Radiation." *George Gamow Symposium, ASP Conference Series* 129 (1997): 84–94.

Zuckerman, Larry. *The Rape of Belgium: The Untold Story of World War I*. New York: New York University Press, 2004.

Index

Page numbers in *italics* indicate figures.